THE WORKOUT BOOK FOR DOGS

A Complete Guide to Fitness Training

Exercises to improve:

- Strength
- Relaxation
- Co-ordination
- Flexibility
- Tactile awareness

Carmen Heritier
& Sandra Rutz

First published in Germany in 2018 by Kynos Verlag

English language edition published in 2022 by First Stone Publishing
an imprint of Westline Publishing Limited
The Old Hen House, St Martin's Farm, Zeals, BA12 6NZ, United Kingdom

ISBN 9781910488614

Printed through Printworks Global Ltd

1 2 3 4 5 6 7 8 9 0

ACKNOWLEDGEMENTS

We want to thank our clients, friends and families for their feedback, practical help and support.

We owe our inspiration to our four-legged friends, who are our daily companions, and to those who are, sadly, no longer with us.

Special thanks to Katrin Häusler for help with the translation.

Contents

How to use this book

This book is for all of those that want to get involved in the world of canine fitness: dog owners, dog trainers, physiotherapists and those who engage in dog sports.

It addresses the physical capabilities of our dogs, and explains how we can work with these to get the best possible results. The fitness programmes we outline, both active and passive, are designed to work on a dog's anatomical strengths and weaknesses, and should be undertaken before commencing more strenuous training.

It is important to bear in mind that tricks and exercises can affect dogs more negatively than positively, if they are executed incorrectly. It is with this in mind that we provide detailed instructions at every stage of learning. However, this cannot – and should not – replace regular check-ups with a physiotherapist or a vet, which should be considered essential.

It is equally important to consider your dog's state of mind. Everything is possible, but nothing is mandatory. Our four-legged friends have bad days, just like us.

This book is also useful as a reference in order to find out which exercises are most suitable for the dog you are working with. We have devised a system of orientation buttons, which will tell you, at a glance, whether an exercised is designed to improve strength, mobility, or co-ordination, whether it addresses sensory issues or is best suited for relaxation. However, we recommend reading the book chronologically as many of the exercises should be seen in relation to each other.

We have also graded the exercises into three categories:
BASIC: Suitable as introductory exercises.
ADVANCED: For dogs who have mastered the basics and are ready for more of a challenge.
PROFESSIONAL: Reserved for dogs who are familiar with the equipment specified, and who have acquired the skills to meet this degree of difficulty.

There is only one thing left for us to say: we hope that you will have a lot of fun with our book. Please do not hesitate to contact us with your feedback and questions.

Sandra Rutz & Carmen Heritier
www.gymnastricks.de

I. THEORETICAL PRINCIPLES

The Elements of Physical Fitness

Strength

Physical strength is defined as the quality, or state, of being strong; it describes the capacity for exertion or endurance. Depending on the type of strength required, and the aim of the training/sport, we can make use of different approaches – but they all require muscular development.

Strength can be categorised as follows:

Maximal strength: The greatest possible strength that can, randomly, be pitted against a resistance. A good example is a dog that pulls on the lead to get closer to something.

Maximum strength requires muscle development, see *In the Gym*, page 198 and *Specific Needs*, page 189.

Speed strength: The ability to use strength at speed. For example, speed strength is required in agility, where the dog needs the power to jump when he reaches the nearside of a fence.

Speed strength can only be accessed via maximal strength. Therefore, muscle development must take place first and can then be followed by training the speed of muscle contraction.

Reactive force: The ability to produce a powerful strength impulse in a muscular stretch-contraction cycle. Reactive force is a special type of speed strength. It involves going directly from one movement to another, so for a human athlete this might be going from a squat to a straight jump.

As it is a type of speed strength, training for maximal strength should precede training for reactive force, for example, plyometric/jump training).

Strength endurance: A dog that persistently pulls on the lead has high strength endurance.

To build up strength endurance, we can either work in the aerobic range of local muscular stamina, or start with muscle development and then work on intramuscular co-ordination.

Since muscular development is required for each of the different types of strength, the exercises in this book will not be differentiated for each category. It is the dog's individual execution of an exercise that will influence the type of strength you train. This is one of the reasons why a concentrated, calm and slow approach to training is essential when you are working on muscular development. If you work out in a gym, you will know that it is the movements that are performed slowly that are the most strenuous. Sloppily performed exercises will not lead to effective training. If you want to train speed strength, train for muscle development first and follow the example of plyometric exercises from human fitness training, i.e. quickly executed exercises of opposing directional movements. However, this should only be attempted after the dog has achieved a satisfactory level of muscular development, also ensuring he has a thorough understanding of the exercises.

Isometry

Isometric exercises involve a special type of strength training. The Greek term 'isometry' describes perfectly what happens in the body:

iso means 'the same'
metric means 'extend'
Therefore isometric means 'consistent' or 'equal length'.

With isometric training, muscles are strained but they don't change their length. It therefore provides the opportunity to train very young or older dogs, as well those recovering from surgery. The fact that no change in muscular length is required means that effective training can be given with minimal demand on other structures in the body. Isometric training works with pressure and counter pressure, which can increase with training.

A good starter exercise is the chin target – "hold" – which is introduced in basics of targeting (see page 30). Isometric exercises, which can be used outside formal training, in a domestic setting, are used throughout this book.

Co-ordination/Body Awareness

For our purpose, we define co-ordination as the ability to put the body in order and to, consciously, use the limbs aligned and geared to each other. Co-ordination is a crucial ability but, unfortunately, it is missing in many dogs. It is, therefore extremely important to teach this skill. Youngsters benefit most particularly from co-ordination training because it teaches them self-awareness. This type of training is also useful for older dogs, helping them to maintain their balance and co-ordination. As co-ordination and body awareness are inextricably linked, we have combined both features in the forthcoming exercises.

Sensomotoric

Sensomotoric, in a classic sense, means the interaction between sensory perception and motor stimulus. In the context of dog fitness training, it concerns the provision of additional tactile stimuli during the performance of exercises. For example, working the dog on a mat with raised bumps instead of a flat surface. This is believed to result in a deeper level of realisation as well as an additional activation, just as with human ergotherapy.

One way to achieve sensomotoric promotion is to work with different surfaces. Creating a sensory pathway using a variety of surfaces is especially valuable for both young and older dogs. Exercises with tactile involvement can be found with the help of the 'Tactile' button.

TACTILE

Mobility

The term mobility, or flexibility, can be linked to physical, as well as mental abilities. While your dog will, hopefully, stay mentally flexible as you gradually build his fitness programme, the focus is on physical mobility, i.e. the ability to get into different postures. A flexible dog has elastic

Hypertrophy: ***Increase of muscle cell volume through exercise***

- Increase of mitochondria *(source of energy)*
- Enlargement of nutrient storage
- Concentration of capillary system
- Higher tensile strength of fascia *(connective tissue)*
- Higher composition of protein
- Increase of strength

Atrophy: ***Decrease of muscle cell volume due to insufficient use***

- Decrease of mitochondria *(source of energy)*
- Reduction of nutrient storage
- Decrease of the concentration of capillary system
- Lower tensile strength of fascia *(connective tissue)*
- Lower composition of protein
- Decrease of strength

Overview of physiological muscular changes that take place as a result of exercise: hypertrophy, and insufficient exercise: atrophy.

muscle fibres that can adapt to strain, which is why flexibility should be an important part of your fitness training.

Endurance

Endurance means the ability to maintain physical exertion over an extended period of time. It also relates to the body's resistance to fatigue, and its regenerative capacity after physical exertion. There are two types of endurance: aerobic and anaerobic:

Aerobic

If aerobic endurance is trained, the body gets its energy from oxidation of the oxygen from the air. Aerobic endurance training promotes enlargement of the cardiac muscle, and a decrease of the resting heart rate. This is because the cardiac muscle becomes more effective, which means that it transports more blood with each heartbeat.

Anaerobic

Anaerobic endurance relates to the body's ability to cope with an oxygen deficit. There comes a point when the exertion is so intense, a sufficient level of oxygen can no longer be obtained from breathing air. This is when the so-called 'excess post-exercise oxygen consumption' kicks in, and the body obtains oxygen through glycolysis (antioxidant process). The level of the excess post-exercise oxygen consumption is trainable.

There are a number of ways to train for endurance (see table, page 10).

The choice of endurance training method depends on what you are trying to achieve with your dog. In most cases, handlers are focusing on the dog's general fitness, in which case, jogging or cycling once a week will suffice. (see Designing a training schedule, page 10).

Principles of Training

In order to improve your dog's fitness, or to maintain it, there are a number of important training principles to bear in mind. A common mistake is to introduce exercises randomly, without paying due attention as to whether they are properly executed. This is often accompanied by inadequate knowledge of muscle training in general. The following section seeks to address these issues.

Fitness training can be broken down into two areas:

1. Factors we can influence
2. Factors we cannot influence.

Factors We Can Influence

Strain

When we are trying to improve fitness, the body must be stimulated to do more than normal, i.e. it must be put under a degree of strain. This is logical, but it is often neglected in the field of dog fitness. In order to maintain a dog's fitness, we have to present consistent challenges; to improve it, the challenges have to be continually enhanced. This is why fitness training for dogs is highly individualised, and why it is impossible to stipulate number, intensity and length of exercises. When we humans want to get fit, we recruit the services of a personal trainer to devise a programme suited to our own particular needs – and it is just the same for dogs.

In order to get a rough idea of your dog's capabilities, and whether he is ready to progress, test to see if he can hold a set position for a duration of 15 seconds, and repeat five times in a row. It may not seem to be much, but it requires a basic level of fitness.

Frequency

This refers to the number of training sessions per week, and is dependent on what you are trying to achieve: decreasing fitness levels, maintaining fitness levels, or improving fitness levels.

Intensity

Within the context of a training session, intensity is a significant factor. In most cases, you can vary the intensity with each repetition, and you can also adjust the way it is executed. For example, you can ask the dog do the exercise on a wobbly surface.

Length

This describes both the length of the training session, and how long the dog holds a position. As a rough guide, a stretch should be held for 10-15 seconds, which is a quite long time for the dog to remain still.

Progression

As already stated, we need to adjust training continuously so that it matches the dog's individual performance level. This is why evaluation is an on-going process. It makes sense to check, every other session, to see if the dog can hold a set position for a duration of 15 seconds, and repeat five times in a row. If the dog is able to this, you can progress the exercise – for example, extending duration or intensifying the exercise by working on a wobbly surface, so that his body is further challenged. If the dog is unable to hold position five times, for 15 seconds, you should simplify the exercise accordingly. If he is just able to keep up with the exercise, you are training at the right level.

Use and Disuse – nothing will come of nothing

If you have broken an arm or a leg, you will know how your muscles weaken through lack of use. It is often stated that muscle wastage occurs after four weeks of inactivity but, in reality, it will be apparent much sooner. This is particularly the case for older dogs, and it is why it is so important to train certain muscle groups continuously.

Specificity – am I really training what I need?

At first glance, 'specificity' is a very obvious training principle. But in the field of fitness training, it is highly complex. Even if you want to focus on certain muscles or body parts, you will, inevitably, bring other areas into play. For example, you cannot confine your training to neck muscles; you can focus on them but, depending on the exercise, you will be involving other parts of the body, such as the shoulder muscles. For this reason, you need to question whether the exercise you have selected is best suited to your training objective. Furthermore, you need to decide whether your aim is to work on strength, flexibility, coordination, or a combination of these factors.

Factors We Cannot Influence

Differences between breeds

One of the factors you cannot change is the anatomy of your dog. Conformation varies hugely between different breeds, mixed breeds, breed types, and between individuals of the same breed.

It is obvious, to the naked eye, that the musculature of a Staffordshire Bull Terrier is very different to that of an Italian Greyhound. But, in fact, the difference is more profound, right down to their muscle fibres.

There are three types of muscles, each with its own specific purpose:

Smooth: Used by the systems of the body, such as the digestive system.

Myocardiun: Also known as the cardiac muscle, it is used to keep the heart pumping.

Voluntary: Also known as skeletal muscle, it differs from the other muscle types as we can influence movement by tensing or relaxing.

The composition of skeletal muscles determines if an individual has the attributes of a strength type or an endurance type. Muscle strands comprise of three different types of fibre that differ in terms of colour, speed of contraction, thickness of contraction, and number of mitochondria, which produce adenosine triphosphate (ATP), the universal energy source for all cells.

Designing a training schedule

Method	Strain	Effect
Duration method *Long lasting strain without breaks – with constant intensity*	***Intensity low to medium (extensive):*** *duration of exertion: up to several hours; aerobic strain* ***Intensity high (intensive):*** *duration of exertion: up to 45 minutes; aerobic/anaerobic strain*	*Basic endurance; strain tolerance; economy of aerobic performance capability; change of muscle fibre (fast twitch to slow twitch, see page 11); fat metabolism basic endurance; strength endurance; long term endurance*
Alternation method *Changing intensity*	*While constantly staying in the range that makes the training effective, intensity changes according to plan or due to changes in terrain*	*Glycogen metabolism, changes in muscle fibres/ mental ability of perseverance and concentration; as with the duration method, adaptability (physiological, mental); recuperation*
Interval method *Change between fairly short strain and relief periods, intervals only for limited relief (incomplete)*	***Intensity low to medium (extensive):*** *duration of exertion up to 10 minutes; aerobic strain* ***Intensity high but not up to a maximum (intensive):*** *duration of exertion mostly up to 60 seconds; aerobic/anaerobic strain*	*Basic endurance; strain tolerance; aerobic performance capability; slow twitch fibre/adaptability; concentration and mobility capability* *Basic endurance and strength endurance in aerobic/anaerobic area of activity/ aerobic and anaerobic performance; lactate tolerance; enlargement of heart volume*
Repetition method *Intensive, fairly short strain periods and long relief periods; short extent*	***Intensity specific to competition:*** *anaerobic train*	*Endurance specific to competition, explosive strength endurance/ anaerobic capacity and performance; lactate tolerance and compensation capability; fast twitch fibre/ mobilisation and perseverance capability at anaerobic conditions*
Competition method *One-off, rarely repeated strain with maximum effort and typical competition performance/ trial competition*	*Duration of competition* *Less than duration of competition* *Longer than duration of competition*	*Complex performance; development of relations between all requirements and their characteristics according to competition*

Muscle fibre type 1 – Slow twitch Marathon runner (> 30 min. strain)

- *Colour: red*
- *Tiring: very slow*
- *Mitochondria: very high number*
- *Contraction: slow*
- *Strength: small*
- *Capacity: thin*
- *Type of sport: long-distance run, bicycle race, triathlon, marathon, iron man (long lasting strain for more than 30 minutes)*

Muscle fibre type 2a – fast twitch Medium distance runner (1-30 minutes strain)

- *Colour: red*
- *Tiring: slow*
- *Mitochondria: high number*
- *Contraction: fairly quick*
- *Strength: medium*
- *Capacity: thick*
- *Type of sport: mediumdistance running, swimming, climbing (lasting strain under 30 minutes)*

Muscle fibre type 2b – fast twitch Weight-lifter (< 60 seconds strain)

- *Colour: white*
- *Tiring: fast*
- *Mitochondria: low number*
- *Contraction: very fast*
- *Strength: high*
- *Capacity: thicker*
- *Type of sport: Sprints, weight-lifting, gymnastics (strain for less than 60 seconds)*

Each muscle contains all three types of muscle fibre, but in varying amounts. This is the result of selective breeding where a breed has been developed to excel at a particular task. For example, a Saluki was bred for fast sprints, and has more fibres of muscle fibre type 2b – so almost twice that of a Dalmatian, who was bred to run alongside carriages over long distances.

Research remains undecided as to whether the quantity of a muscle fibre type can be influenced by training, i.e. changing from slow twitch to fast twitch. However, it is definitely possible to improve the performance of each muscle fibre type. A Staffordshire Bull Terrier will never become a Siberian Husky just because of muscle fibres, but that cannot be an excuse for a lack of fitness.

Cellular adaptation – physiological factors regarding training

When environmental influences affect the body, it reacts or even adapts. For example, if we are in contact with cold air, our natural reaction is to get goose bumps. With fitness training, we want to trigger the same sort of natural reactions/adaptations. To achieve this, a cellular adaptation needs to take place in order to prepare the body for new tasks. So, over a period of time, structures within the cell are re-programmed. Think about running a marathon. As an average athlete, you would never contemplate competing in a marathon next week. Your body needs time to adjust to the challenges that lie ahead. This is exactly the same for our dogs.

In the area of strength training, we need to leave a gap of at least 48 hours between training sessions. This is the time it takes for the cells to adjust to modified strain. Strength training twice a week, with a break of 48 hours, is considered to be the ideal. This should take place over a 12-week period, which allows the body to adjust completely to the new tasks.

Research has proved that attempts to shorten the 48-hour break between training sessions has a detrimental effect. The dog's performance starts to plateau, and deteriorates. Therefore, the 48-hour rule should be considered compulsory.

Fitness Tests

In order to assess which exercises are suitable for the dog we are working with, we need to conduct a series of fitness tests. These tests also help to identify weaknesses, especially if the dog competes in high performance sports. In addition, we can work out a training strategy, for example if the muscles on one side are weaker than on the other. This is particularly important for sports dogs who may rely heavily on a particular set of muscles, and will, therefore require compensatory exercises. For example, if a dog does a lot of heelwork, holding his head high, you would compensate by giving him exercises that encourage him to look downwards or underneath his belly. In contrast, a tracking dog, with his nose to the ground, is putting a lot of weight of his front legs. To compensate for this, you would work on exercises where he has to place his front feet on an elevation, which means he has to put more weight on his hind limbs.

Fitness tests are divided between active and passive. Passive tests involve the dog being checked physio-therapeutically to detect muscle tension and possible dysfunction. If there are noticeable problems, you would need to consult a specialist before progressing to active fitness tests. Naturally, dogs are bred for different tasks and are engaged in different types of sports. Sled dogs are likely to have good endurance but may be weaker in co-ordination, as this is not integral to their task. In contrast a dog who does canine freestyle, will be strong in body perception and body control.

The following tests are not directed at the specific levels of fitness required for each area of the body; they are intended as a rough indication of the fitness of your dog.

The dog should stand straight and parallel while you stroke downwards, assessing the femoral muscles/thighs.

Passive Tests – Muscle Status

a) Muscle Test: Thighs

Position yourself behind the dog and simultaneously stroke down each of the femoral muscles, checking for asymmetry. Use both hands so you can assess differences between the muscles.

We now know that measuring thighs is unreliable; atrophies are picked up at a late stage and it is unclear which muscles are affected. It is therefore much more effective to conduct this test through touch and feel. The thigh should feel flat on one side and arched on the other side.

Please note: for the passive tests, the dog has to stand straight, with his legs parallel and facing forward. To achieve this, will need to recruit a helper to assist you.

With the dog standing straight, encompass each thigh with your hands, feeling for differences between the muscles.

b) Muscle Test: Hip extensor

Place both hands (with fingers closed) on the dog's croup. You should not feel a dent but, rather, a smooth, upward curve. This result often coincides with the muscle status of the thighs. For example, you will be able to detect whether the whole limb has been favoured, either consciously, or unconsciously, over an extended period of time.

With the dog standing straight, place both hands on the croup and assess the differences between the muscles on both sides.

c) Muscle test: Upper arm

Position yourself behind the dog and, place each of your hands on the upper front legs (musculus triceps brachii). Palpate the muscles of the front upper leg area (musculus biceps brachii), watching out for asymmetry, but also increased muscle tension. Asymmetry of the front limbs often appears diagonally to asymmetry of the hind limbs. For example, the back right shows less strain; the front left shows more strain.

The helper needs to ensure that the dog's front legs are parallel. Use your hands to check the differences between the upper leg muscles.

Now touch the back of the upper leg.

Hands are placed parallel to the spine. Step by step, check the muscles next to the spine for differences.

d) Muscle test: Spine

Place both hands (with fingers closed) parallel to, and alongside, the spine. Palpate the muscles next to the spinous processes,(the bony projections that project posteriorly from each vertebra), starting directly behind the shoulder blades. If one side feels more prominent, even partially, it is probably better muscled than the other side. If muscle hardening has occurred, the dog would duck away at punctual pressure, which involves stroking over the muscle, and then gently probing with a finger tip.

Overburdening Test – Forehand

During the course of a lifetime, many dogs put too much stress on the forehand. This could be the result of an orthopaedic/neurological problem in the hindquarters, the dog may have incorrect angulation of the shoulder joints, hindering movement, or he may have been exercised inappropriately – for example, playing excessive games of fetch. Older dogs will often show signs of overburdening of the forehand. These include walking with the head held low, a wide front, an exaggerated torso and hyper-extended paws.

To test for this condition use both index fingers, simultaneously, to apply pressure to the wide back muscle towards the centre of the body. The best place to do this is directly behind the shoulder blade. If there is overburdening, the dog will immediately duck away, or the muscles will react by twitching. In this case, you should get your dog checked, and then choose exercises that encourage the dog to put more weight on his hind limbs. *See Exercises to train body awareness and co-ordination of the hind limbs, page 170.*

Active Test – Mobility of the Spine

a) Motion of the cervical spine (neck)

This involves testing the vertical mobility – upwards and downwards – towards the sternum, and sideways towards the shoulder blades. You can use a treat to lure the dog into the appropriate position. He can be sitting, or standing. If one side does not bend easily, or the motion falls short, upwards or downwards, you will have to focus on this in your exercise programme.

Use your index fingers to apply pressure on both sides, directly behind the shoulder blade. Then continue to apply pressure, working from top to bottom of the shoulder blade. If there is evidence of overburdening, you will notice a twitch.

To test the mobility of the cervical spine in an upward direction, encourage the dog to stretch his neck upwards as far as possible.

Using a treat, encourage the dog to move his head towards his flank. Here we have used dog-steppers to keep him in position.

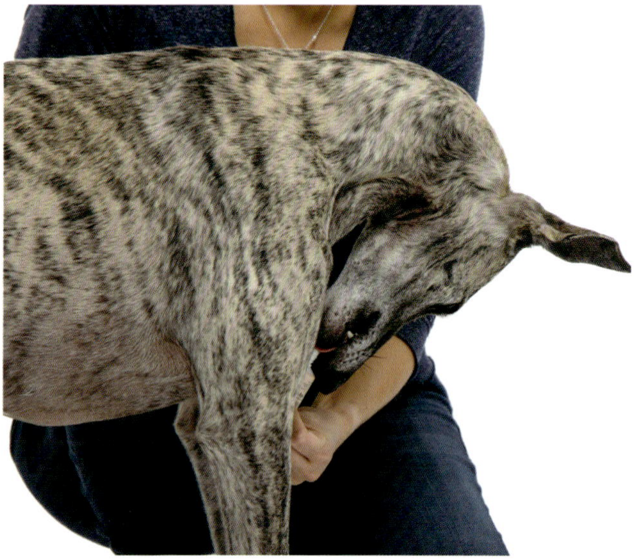

Now move the dog's head towards his sternum.

To test the upper cervical spine, move the dog's head towards his shoulder blade.

b) Movement of the entire spine to both sides, and forwards.

With the dog in a standing position, lure his head sideways, towards the knee. Repeat on the other side – you are looking for equal movement in both directions. Now lure the dog so that he is looking underneath his belly. You will need to perform this exercise very slowly to allow the dog to maintain his balance.

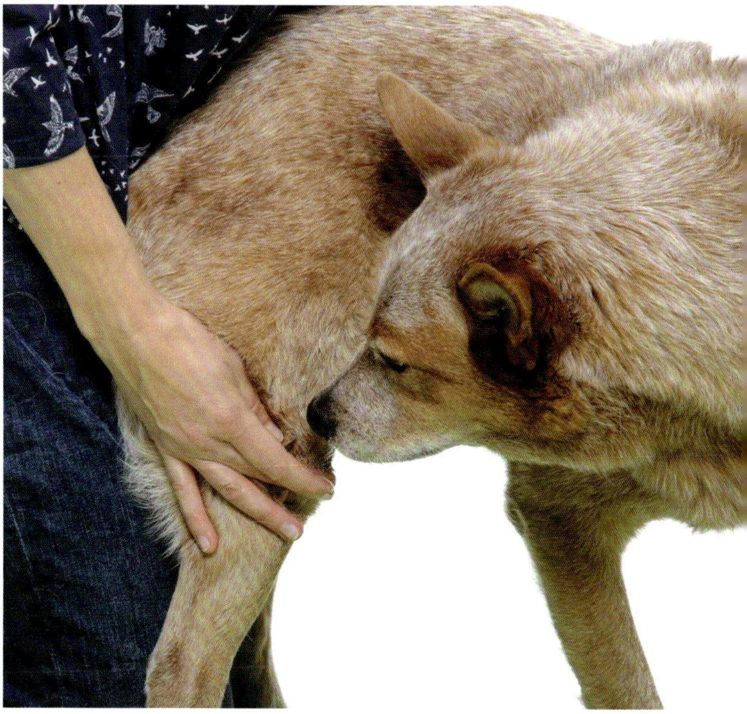

The dog's head is moved towards the knee.

Lure the dog's head through the front legs. In order to achieve a slower execution, we worked with a Post-it as a target, and stationed the dog on a dog stepper.

Active Test – Holding Positions

a) Torso-forehand stability

Position the dog's hind legs on a platform that is roughly as high as his elbows. The backline should be straight, and slant towards the floor; the elbows can be slightly bent. The dog should hold this position for 15 seconds, five times in a row. Between the repetitions, release the position briefly and then position the dog again. With this exercise, there should be no signs of discomfort, which may be seen as evasion, muscle twitching, or panting.

b) Standing on three legs

In order to test the stability of the dog's hind legs, you will need to elevate his front legs at elbow height. Lift each of his front feet, in turn, holding the position for 15 seconds, five times in a row. If the dog pulls his leg away, do not attempt to hold on to it. The hind legs should not move during this exercise, but make sure you allow the dog to balance himself before holding the position.

Position the dog with his hind legs elevated on a platform; he should stand parallel, both front and back, and hold the position.

Equipment: Ikea stool for children.

Position the dog with his front legs elevated, and lift each foot in turn to transition the weight to the hind legs.

Equipment: Ikea stool for children.

Active Test – Power Boost From The Hind Legs

The dog should be slightly elevated (approximate height of his hocks) and is lured from a sit to a stand. The exercise should be repeated three times with ease. Often, a dog has learnt that "sit" means keep all four feet on the ground, so a refusal does not necessarily mean he is experiencing difficulty.

If the dog has understood the exercise on the small elevation, you can try it with a slightly higher elevation. As he goes from sit to stand, there should be no compensating movement of the hind legs. If in doubt, repeat the exercise three to four times to rule out the possibility that the movement was incidental and, rather, it is indicative of a problem.

This exercise puts a lot of strain on the body so take care with older dogs. Dogs with cauda equina syndrome (progressive arthritis in the joint at the bottom of the spine) should not be subjected to this exercise.

Lure the dog from a sit to a stand. Here, you can see compensating movement of the left hind. In fact, you want the dog to stand parallel.

Equipment: Togu Aero Step One.

Designing a Fitness Programme

When you are planning a fitness programme, you should schedule two to three sessions, per week, with a break of 24 to 48 hours between sessions, depending on the dog's additional activities. If you follow this basic set up, it will usually be 6-8 weeks before you see a noticeable improvement. During this period of time, you should increase the level of difficulty, the number of repetitions for each exercise, and the time frame of the training. However, a fitness training programme usually takes about 12 weeks to complete.

Make sure the dog is well rested before the start of each training session so that his muscles are not already tired; and give him the opportunity to toilet. In terms of the dog's exercise regimen, a training session replaces a walk.

Regeneration

Between training sessions, the body needs to restore energy, dissipate metabolic products and build new structures. The longer the training session lasts, the longer the body needs in order to recover. If you scheduled a training session every day – possibly for the same group of muscles – the body would have no time to recover, and training would be ineffective.

There is a clear difference between training in a group or training alone in your own training facility at home. A group session usually lasts 45-60 minutes and includes a 10-minute warm-up, a training session with a range of targets (beginning with strength and co-ordination), ending with a 5-minute cool-down. During the session, the dogs take turns between exercise and rest breaks. If you train alone, make sure you plan sufficient rest breaks for your dog. These breaks will depend on his levels of fitness and performance, therefore, the table (below) can only serve as an guide. Training schedules are dynamic and change progressively. You will soon get a feeling of how to work your dog once you have started training.

Build-up over a period of six weeks with two to three training sessions each week, and four to five exercises per session. Duration of each session (without breaks): 8-20 minutes.

	Repetitions / Exercises	*Rest breaks*	*Length of training without breaks*
Week 1	*3-5 reps per exercise/2 min.*	*1 minute per exercise*	*8-10 minutes*
Week 2	*3-5 reps per exercise/2 min.*	*1 minute per exercise*	*8-10 minutes*
Week 3	*4-6 reps per exercise/3 min.*	*30 sec.*	*12-15 minutes*
Week 4	*4-6 reps per exercise/3 min.*	*after two exercises*	*12-15 minutes*
Week 5	*5-8 reps per exercise/4 min.*	*after half of the training time: 1 minute*	*16-20 minutes*
Week 6	*5-8 reps per exercise/4 min.*	*after half of the training time: 1 minute*	*> 20 minutes*

If you notice signs of the dog tiring during an exercise, or towards the end of the training session, you should end the exercise or the training. Finish with an easy target exercise so the dog does not have a negative experience. At the next session, reduce the exercise, or the entire session, to 70 per cent and then start to increase gradually.

When working on a training programme, with the exercises presented in this book, you need to be aware of the fact that you never train one single group of muscles but, rather, several muscles and groups of muscles at the same time. However, for the sake of simplicity, we have summarised exercises that can benefit certain muscles if there is a weakness, or if you want to focus on a certain part of the body. Our classification is guided by what we believe is relevant for training sports dogs. However, the pelvic floor has never been addressed before, so we will break new ground in this area.

Please note: Training is only stimulation. Adaptation of the body (super-compensation) occurs during rest periods.

The Dog's Muscle Groups and their Functions

1. Nuchal ligament (Ligamentum nuchae)

A long, slightly stretchable bowstring ligament, that reaches from the axis to the 4th dorsal vertebra. It is attached to the 3rd, 4th, 5th and 6th cervical vertrebra and the first 4 dorsal vertrebra.

Function: It holds neck and head, controls the movement of the head and serves as a connecting point of the muscles that produce the forward movement of the forelimb, as well as keeping the shoulder blades in position during the backward movement of the forelimb.

2. Top shoulder muscle, trapezius muscle (musculus trapezius) (in green)

This muscle consists of two parts, which are attached to the spina scapulae: one at its front, and one at its back. The front part is also connected to the nuchal ligament (beginning from the 3rd or 4th cervical vertebra) and the first dorsal vertebrae. The back part is also attached to the 2nd and to the 9th (or 10th) dorsal vertebrae.

Function: It keeps the tip of the shoulder blade in position, controls the movement of the upper part of the shoulder blade and is involved in the lifting and forward movement of the forelimb.

3. Undermost shoulder muscle (musculus serratus) (in blue)

This muscle also consists of two parts, which are attached to the lower part of the spina scapulae, one is attached to its front, and one at its back. (The front part is also connected to the nuchal ligament (beginning from the 3rd cervical vertrebra).The back part is attached to the ribs (to the 7th or 8th rib).

Function: Controls the movement of the lower part of the shoulder blade, takes part in moving the forelimb and carrying the body.

Any activities of the musculus serratus and musculus trapezius are closely connected to each other: 2a and 3a, as well as 2b and 3b always work together.

4. Head-neck-arm muscle (musculus brachiocephalicus) (in green)

A long muscle that reaches from the base of the skull to the head of the humerus.

Function: If the nuchal ligament is stretched, and therefore the head is carried upwards, this muscle pulls up the forelimb by contracting itself. If the nuchal ligament is relaxed, thus contracting the musculus brachiocephalicus, the head is pulled downwards.

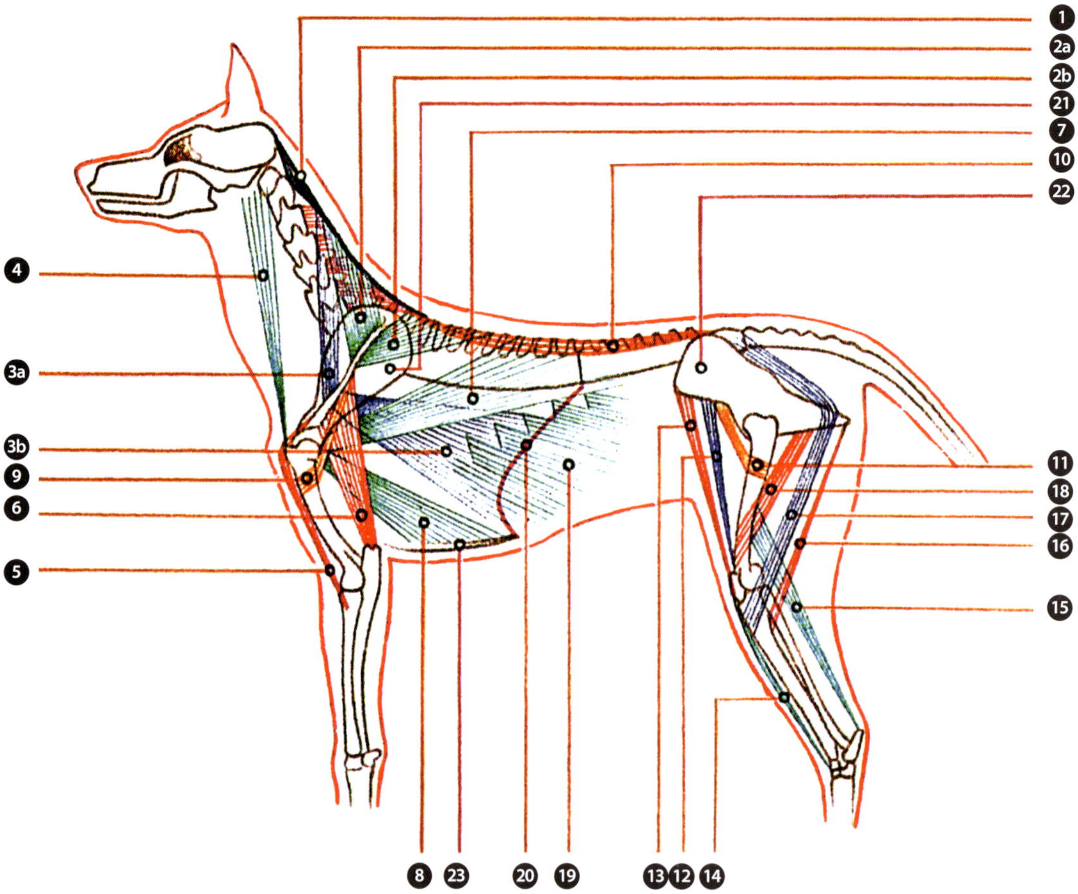

5. *Flexors of the elbow joint (musculus biceps brachii)(in red)*

A thin muscle that reaches from the lower part of the spina scapula in front of the head of the humerus and along the humerus towards the spoke.

Function: Serves as the flexor of the elbow joint and keeps the head of the humerus in position.

6. *Extensors of the elbow joint (musculus triceps brachii) (in red)*

A strong muscle that is attached to the spina scapula and the tip of the elbow.

Function: Extends the elbow joint and thereby the forelimb.

7. *Wide back muscle, pectoralis major (musculus latissimus dorsi) (in green)*

This is a broad muscle that is attached, at its back, to a big aponeurosis at the loins and at the last four dorsal vertrebrae.

Function: Brings back the shoulder in collaboration with the breast muscle.

8. *Breast muscle (musculus pectoralis) (in green)*

This muscle is attached to the breast bone (sternum) and the underpart of the ribs on one side, and to the head of the humerus on the other side.

Function: Brings back the shoulder in collaboration with the wide back muscle.

9. *Deltoid muscle (musculus deltoideus) (in orange)*

The deltoid muscle is attached to the spina scapula and the upper part of the humerus.

Function: flexes the deltoid.

10. Long back muscle (musculus longissimus) (in orange)

This very important muscle is attached to all dorsal vertebrae from the loins to the neck.

Function: Forms the base of the nuchal ligament. Keeps the spine in place, allowing it to move, although other muscles are influencing it, causes the thrust through the spine, carries a great amount of the body.

11. Flexors of the hip joint (in orange)

This muscle is attached to the hip bones and the upper part of the humerus.

Function: Flexes the hip joint.

12. Extensors of the thigh (in blue)

One end is attached to the hip bone and the other one at the lower part of the humerus.

Function: Pulls the humerus and, thereby, the hindlimb, forwards.

13. Extensors of the knee joint (in red)

Reaches from the hip bones to the head of the shinbone.

Function: Keeps the knee straight during the forward movement of the hindlimb.

14. Flexors of the ankle joint (in green)

This muscle is attached to the shinbone and the ankle joint.

Function: Flexes the ankle joint.

15. Extensors of the ankle joint (musculus gastrocnemius) (in green)

This muscle is attached to the underpart of the humerus and linked to the heel by the Achille's tendon.

Function: Extends the ankle joint.

16. Flexors of the knee joint (musculus biceps femoris) (in red)

This muscle is attached to the ischius and the shinbone.

Function: It can flex the knee joint if the musculus semitendinosus is not working.

17.– 18. Sprint muscles, musculus semitendinosus

These are two strong muscles. Number 18 (in red) is attached to the hip joint just above the knee joint. Number 17 (in blue) comes from the sacrum and reaches over the ischium to the head of the shinbone. Both muscles reach around the knee joint, keeping it extended during the forward and backward movement of the leg.

Function: If the leg moves forward and the foot makes contact with the ground, the muscles serve as shock absorbers. They then keep the knee joint extended, whereas the extensor of the ankle joint keeps the ankle joint extended. After the foot makes contact with the ground, these muscles contract, and thereby provide the thrust that is needed to move the body forward. The knee and ankle joint have to be extended in order to achieve the greatest thrust.

19. Slant abdominal muscle (in green)

This muscle is attached to the ribs and to the aponeurosis of the lower belly.

Function: Controls the dog's breathing.

20. Diaphragm

The diaphragm is also called the pelvic floor and is a connective muscle tissue that is located between the ribs. It is attached to the 7th rib and then stretched back to the end of the breast bone. At its top, it is attached to the loins.

Flunction: Controls the dog's breathing.

21.Shoulder blade (scapula)
22.Hip bone (pelvis)
23. Breast bone (sternum)

Targeted Training

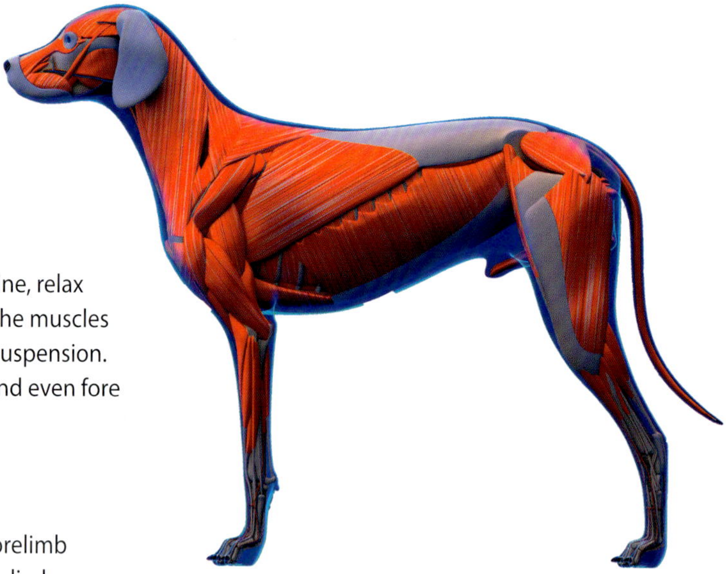

Neck muscles and shoulders

These exercises mobilise the cervical spine, relax the surrounding structures, or activate the muscles of the cervical spine, core muscles and suspension. Depending on the exercise, shoulders and even fore limbs will be involved.

Muscles:

- All movable muscles of the head.
- Musculus brachiocephalicus (head-forelimb muscle) – forwards movement of fore limbs, muscle for locking in position, sideways bend of the neck, and pulling down of the head.
- Musculus splenius capatis (long head muscle) – extensor and for lifting, lateral bending of neck and head.
- Musculus longus colli (neck flexor) – flexor for cervical spine.
- Musculi scalene (middle and back muscles for holding the ribs) – flexor, lateral bending and muscle to help with breathing.
- Musculus longissimus (long back, neck and head muscle) for lifting, extensor, muscle for locking in position and lateral bending.
- Musculus spinalis et semispinalis thoracis et cervicis – for lifting, muscle for locking in position, extensor spine and muscle for straightening the torso.
- Musculus transversospinalis – lifting, locking device, extensor and lateral bending of head and neck.
- Musculi multifidi (feathered muscles) – muscles for locking in position, turning and lifting.
- Musculus iliocostalis (stabilises and straightens the spine) – does not concern the neck as such, but active while lateral bending.
- Pectoral girdle (see fore limbs).
- Core muscles (see fore limbs).
- Lateral abduction muscles for lateral bending.

Fore limbs

In this section, you can find all the exercises that train the upper front leg area as well as the pectoral girdle and its suspension. They are mostly strengthening exercises or are designed to improve the extension of the forelimbs.

Muscles:

- Musculus trapezius (trapezius muscle): stabilisation and forward movement of the shoulder blade.
- Musculus brachiocephalicus (head-forelimb muscle): forward movement of forelimbs, muscle for locking in position, sideways bending of neck and pulling down of the head.
- Musculus amotransversarius (shoulder-head muscle): forwards movement of limbs, muscle for sideways movement of the head and neck.
- Musculus latissimus dorsi (broad back muscle): pulling back of limbs.
- Mm. pectorales superficiales (pectoral muscles): muscle for holding the torso, for bending sideways, extensor and muscle for locking shoulder joint in place, important when turning around on the fore hand. Subject to atrophy in older age.
- Musculus rhomboideus (rhomboid muscle): stabilisation, lifts the shoulder, forward

movement of fore limbs and neck, and locking in place of shoulder blade.

- Musculus serratur ventralis (ventral ridged muscle): stabilisation in stance phase (gait: foot in contact with the ground), muscle to help with breathing, suspension stance (gait: loading response), forward and backward movement of limbs in general, sideways bending of the neck.
- Musculus supraspinatus (upper bone muscle): extensor of the shoulder.
- Musculus infraspinatus (lower bone muscle): lateral stabilisation of shoulder joint, muscle for bending the shoulder and for slight external rotation.
- Musculus deltoideus (deltoid muscle): stabilisation of shoulder joint.
- Musculus biceps brachii (biceps muscle): extensor of shoulder joint and bending of elbow joint.
- Musculus brachialis (brachialis muscle): bending of elbow joint and muscle to rotate limb outwards.
- Musculus triceps brachii (triceps muscle): extensor of elbow, bending of shoulder joint, important muscle for stance phase (gait).
- Musculus tensor fasciae antebrachii (antebrachial fascia tensor): extensor of elbow joint

Front femoral/front thigh muscles

It is not necessary to strengthen the front femoral muscles as often as the side femoral muscles. Still, the exercises intended for these muscles are useful, even if it is only for compensation. We often train the quadriceps muscle group with dogs that suffer from luxating patella, a condition involving a loose kneecap, (see page 191).

Muscles:

- Musculus quadriceps femoris (quadriceps knee extensor): muscle for extension of the knee, fixation of limbs and takes part in bending the hip joint.

Muscles that are additionally trained (depending on exercise) are:

- M. sartorius (sartorious muscle).
- M. tensor fasciae latae (extensor of crural fascia).
- M. iliopsoas (iliolumbar muscle).

All of these muscles play a significant role in swinging forward the hind limbs.

Exterior femoral/outer thigh muscles and gluteal muscles

This area of the body often requires work because many weaknesses can be found here. We have summarised the exterior femoral muscles and the gluteal muscles in one category as, on the one hand, they are mutually affected and on the other hand, they are easy to train together and combine.

Muscles:

- Musculus biceps femoris (biceps femoris muscle): with its two parts, it plays a significant role in the stance phase (gait: 30-50 per cent), sideways pulling, extensor of the hip, the knee, bending of the knee and extensor of the ankle joint.
- Gluteal muscles (croup muscles) are mainly responsible for extending the hip, stancing forward and taking part in abduction (pulling forward).
- Adductors and the hamstrings are additionally trained.
- The exterior femoral muscles are the muscles that are most affected by muscle atrophies. Generally the adductors are tense when there are atrophies in the exterior femoral muscles, so lateral exercises work best.

Relaxation of back muscles

Relaxation of the back muscles should always be planned for the end of your training session, as the muscles go into standby mode and, as a result, there is a higher risk of injuries if you do strengthening exercises subsequently (low body tension).

Back muscles:

Flexing, extending, sideways bending and fixating spine, power transmission from hind legs to front legs.

The most important muscles of the spine:

- M.iliocostalis (spans from the ribs to the spine and stabilises this region)
- M.longissimus (long back, neck and head muscle)
- Mm. multifidi (feathered muscles)
- Mm. rotatores (rotator)

Abdominals:
Supports motion, works in opposition to the back muscles, can tense segmental-reflexal (e.g. if back muscles are tense), supports breathing, defecation, urination and labour.

Iliopsoas muscle: Moving forward of the swinging leg in the swing phase, hip flexor (turns knee outwards at the same time), works in opposition to the gluteal muscles.

Activation of back muscles

This is important for sports dogs, as the back has to play a significant role in compensation. First of all, check for a difference in both sides and then build up the back muscles completely, or with focus on one side, which means doing more repetitions for the weaker side. We often find a difference in the strength of both sides in dogs that do a lot of heelwork only on one side, or dogs that get up momentum from the lumbar vertebra one-sidedly.

Pelvic floor

Here we will follow an approach to strengthen the pelvic floor in order to avoid, or counteract, bladder weakness. There is yet to be empirical evidence for the success of this practice, but we believe it is worthwhile.

The pelvic floor is the fascial-muscular basis of the pelvic cavity. Its task is to secure the position of pelvic organs and to support the urethra and the sphincter of the urinary bladder. It also supports the closure of the anus. The pelvic floor is segmented in diaphragm anale, perineum and the diaphragm urigenitale.

Body awareness/coordination of the hind limbs

Body awareness and co-ordination issues most often apply to youngsters that are still growing, to taller dogs, and those dogs that know only one way to go (i.e. at top speed) so they are unable to concentrate and perform exercises calmly. In addition, these exercises are suitable for dogs that weight-bear on their fore hand, for example, if there is a problem with the hind legs, or with senior dogs who tend to put most weight on their front legs.

The gluteal muscle, front and back femoral muscles, M. gastrocnemius (calf muscle) and, with lateral stepping, also M. biceps femoris (biceps muscle) are all involved here.

Depending on the dog's gait as he walks backwards, i.e. how narrowly he steps, back muscles and abdominals, as well as M. iliosoas (iliopsoas muscle) are exercised.

Training Know-how

There are a number of training strategies and exercises to teach your dog before you start his fitness programme. They will enhance his understanding, so training sessions will be both effective and stress free.

Clicker training

Clicker training is invaluable when it comes to precise, purposeful training. All exercises in this book will be developed through clicker training, unless luring with treats makes more sense for some, specific exercises.

Clicker training does not mean that you have to work with a clicker in your hand; you can also use a marker word. Obviously, you must train the dog so he builds the appropriate association with your marker word.

You can use a simple "yes", or you may prefer to use a term of praise, "good', for example. However, it is important that you choose a marker word that you only use in a training context so that it has significance for your dog.

We define a 'click' as meaning two things: "What you are doing at this moment is correct" and "I promise you a treat". The click acts as an affirmation of the dog's behaviour, even if it is very complex. This is very important for effective fitness training. It also tells the dog that he will, indeed, get a reward, which will generally be a treat.

We have described the exercises in a series of steps, but this should be taken as a guide. Your dog is in the best position to inform you how to progress through an exercise. We think that it is crucial that you proceed step-by-step. As a consequence, you will not only mentally occupy and engage your dog, but you will also give his body time to adapt to the tasks and challenges he faces.

Theoretically, you could work through the exercises outlined in this book by luring with food. But the dog will not process what he is doing, and the results will not be as effective.

You could also use free shaping (the dog offers behaviours until you reach the desired goal) as a means of training your dog to perform the exercises. If this is your preferred option, we have provided starting points for each exercise, with additional assistance if the dog is struggling to find which direction to take.

Luring with food

For a few of the exercises, it is preferable to lure the dog with food. However, these are the exceptions. Luring with food only makes sense when the dog has to change position without moving from the spot, for example, sit-stand down. In this case, active, autonomous thinking, which is crucial for most of the exercises, is counter-productive because it usually involves movement. We have marked the exercises where luring, as opposed to step-by-step training, is recommended.

Varying the reward

In dog training, rewards should always be rich in variety, so that training is a highlight for the dog and is always surprising and new. In fitness training, reward options are often restricted due to the nature of what we are trying to achieve. For example, using a toy as a reward would be counter productive, in many cases, because it leads to the dog moving from position. Therefore, rewarding with food is usually the only viable option. However, if your dog is toy mad, you could include a game of tug at the end of the session.

The value and significance of the reward is very important in fitness training, and this works best if you use foods of different value. For example, kibble would represent a 'normal' reward and cheese would be a higher value treat. In many instances, you need a treat to get the dog to the intended starting position, or to do a preparatory exercise, prior to your targeted training. In both these scenarios, it is sufficient to reward with kibble – as long as the dog rates it as a reward – so that he remains calm and focussed. If the dog goes on to do something special – a really great execution of the exercise – you can reward with the higher value cheese. It is worth noting that, in this situation, providing bigger pieces of food is beneficial, as chewing can be a calming activity for dogs who are more highly aroused.

Position for rewarding

Click for action, feed for position

Among those who work with clicker training, there is a saying: 'Click for action, feed for position'. The thinking behind it is that you use a click to reinforce an action/behaviour, and then feed the dog in the desired position. For example, if you click your dog for sitting down – bottom moves down and touches the floor – you should also feed him in the sitting position. If you feed him when he is standing – after you have clicked for sitting down – he might learn to hold the position briefly, and then jump up to get the treat.

This is especially relevant if you are working with specific parts of the body, or muscle groups, as the dog needs to keep position for an extended period of time. You know this from your own fitness training: getting into position is usually no big deal; it only becomes strenuous when you hold the position for a longer period of time.

Feeding in position is particularly important when training the different steps of an exercise because each end position is the starting position for the next step. With some exercises, this can be challenging because you have to be quick with handling and rewarding. Luckily, our dogs are very clever and will, generally, understand what their task is.

Take *click for action, feed for position* as your personal challenge while practising the exercises outlined in this book and, in this way, it will not only be the dog who learns.

Alternative position for rewarding

For some of the exercises, you need to differentiate the position for rewarding from the position of execution. Your dog can go to get the food and then return to you, so that a fluent execution is possible. For this purpose, your dog should be familiar with picking up food from the floor that you tossed for him. It can be helpful to work with a positioned food bowl.

If the concept of differentiating positions of execution and rewarding is advisable in the forthcoming exercises, it will be clearly indicated in the set-up instructions.

Basics of target training

For fitness training, targets are often essential because they allow the dog to go to a specific point independently and autonomously. This mean you can be more flexible in your own position, in relation to the dog, enabling a better visual angle.

Take time to work on your target training, as the benefits are immense. It prepares the dog, mentally, for the planned exercise and, in group sessions, it is the perfect warm up exercise to prepare for the work ahead.

Hand target

For many exercises it is helpful if the dog learns a hand target, where he touches the palm of your hand with his nose. This allows you to use the hand target in the same way as if you were luring with food. The advantage is that the dog processes what he is doing rather than blindly following the food. In addition, it makes your job easier when you are working with a clicker and treats.

1. For the set-up, make sure you have a clicker and treats at the ready.
2. Hold the palm of your hand in front of the dog's nose. Tense your hand to make it more interesting. As soon as the dog interacts with your hand, click and reward.
3. Repeat a couple of times. Make sure you present your hand in different positions right from the start.
4. The dog will quickly understand that when you present your hand, it is a signal to touch it. However, it is important to introduce a verbal cue as it will give you greater flexibility when you are training.

Chin/hold target

A chin target, or hold target, is the perfect way to teach your dog to hold position, thus adding duration to an exercise. He also learns to put weight in your hand.

1. You need to be at eye level with your dog, which depending on the size of dog, can be a challenge. It may help if you work with the dog on a sofa. Ideally, he should be in a standing position for this exercise.
2. Show the dog you have some high value treats, but don't give him one yet.

3. Face the dog and cup your inside hand to provide a place for his muzzle, or his whole head (depending on his size). Hold a treat in your other hand.

4. Now place your cupped hand directly between the dog's muzzle and your outside hand, which is holding the treat. This means that dog only has to move slightly towards the treat and he will, automatically, slide his muzzle into your cupped hand. You have to vary the distance, and position of the treat, until you find the right position. Click and reward the moment the muzzle makes

Vary the position of the hand target so the dog learns to generalise, and will offer the behaviour in all situations.

contact with your hand, making sure it is lying flat in your hand.

5. Practise a few times, clicking and rewarding each time the dog places his muzzle in your cupped hand.
6. You are now looking for the dog to push his muzzle into your hand, and to bear down so he is putting some weight into it. To achieve this, repeat the previous step but, this time, move the hand holding the treat downwards. This will involve a slight delay in giving the treat. This manoeuvre requires some skill so work on it in small steps. The aim is for the dog to keep his muzzle in your hand, but to look downwards, which means he will press his muzzle into your hand. Click the moment he does this and give him a treat as he holds the position. Repeat several times until your dog understood the concept.
7. Progress by offering your cupped hand lower and lower. As an alternative, you can also only start moving your hand downwards when the dog has already fully placed his muzzle in it.
8. Now, try to lengthen the period of time in which the dog stays in position, pressing down on your hand. Most dogs find this aspect relatively easy.
9. Introduce a verbal cue and fade the treat gradually by using it on a random basis.

The set-up for the chin target, which involves luring with a treat.

This is what you want the chin target to look like.

Nose target

For some of the exercises in this book, you will also need a nose target. It will help if you use a Post-it/sticky note to teach this exercise. The set-up is similar to that used for the hand target.

1. Offer the dog a Post-it at nose level and click when he first shows interest.

2. Repeat several times, holding the Post-it in different positions.

3. The aim is for the dog to approach and touch the Post-it reliably, and autonomously, in any imaginable position.

4. The visual target is generally enough of a clue to trigger the behaviour. However, if you want to introduce a verbal cue it has to be different from your cues for a hand target (see page 30) or a paw target (see page 34).

You can use any kind of self-adhesive paper to teach a nose target.

Paw target

For many of the exercises in this book, you will need a reliable paw target. You will need to teach a separate target for the front feet and for the back feet. Make sure you use targets that look different, so the dog understands which feet he needs to use. For example, you could use a rubber mat for the front feet, and a firm cushion for the back feet. You will also need separate verbal cues to differentiate between them.

1. Prepare clicker and treats.
2. Hold your target (e.g. a rubber mat) and show it to the dog: "look what I've got…"
3. Put the target down on the ground and click for the dog's first interest in it. Toss the treat away from the target so the dog has to move to get his reward.
4. As he goes to get the treat, pick up the target. When the dog returns to you, put the target on the ground. Now, only click if the dog touches the target with his front feet and toss the treat again. This time, leave the target on the ground.
5. Repeat several times until the dog positions both front feet on the target. Work on duration, giving the dog several treats to keep him in position.
6. Now introduce a verbal cue for front feet on the target.

The target for the back feet can be practised when you are working on backwards movement (see page 40). Remember to introduce a different verbal cue.

Feed your dog when he is position, on the target, to build a positive association.

Training Tools

In this book, we use a variety of training aids for the exercises we teach. You can buy specialist equipment or, in some cases, you can do down the DIY route. Safety is the most important consideration so you need to avoid sharp edges, slippery materials, or any other potential hazards. All equipment should be checked regularly.

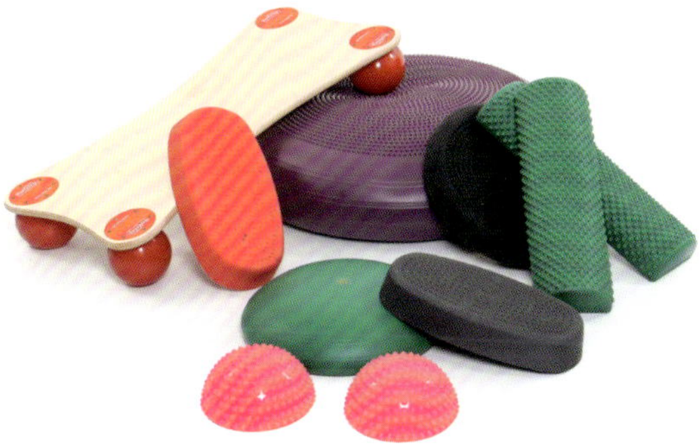

Wobble cushion

For many exercises, you will need a wobbly/unsteady base, which can take the form of a wobble board or a wobble cushion. If you are using an inflatable wobble cushion, you need to consider how much air you pump into it. Unfortunately, it's not as easy as saying: "the more air you pump in, the wobblier it will be". This is because the weight of the dog working with the equipment also matters.

For example, a Bernese Mountain Dog would find a moderately inflated cushion fairly easy because his weight compresses the cushion and creates a steady base to stand on. In contrast, a Papillon would find the same cushion very challenging. A dog of this size would find it easier to balance on a fully filled and firmly inflated cushion, which would, in turn, create a challenge for a heavy dog. You may find it helpful to work with several inflatable cushions, with different degrees of hardness and different levels of inflation, so it is easy to adapt to the requirements of your dog.

Correct Size

It is important that all training equipment is the correct size for your dog. He should be able to stand on it with ease, unless narrow standing is required for an exercise, such as *mountain goat* (see page 57). If you are worried about size, for example if you are using a peanut or a platform, take a photo with your dog in position, and have a good look at it. As an alternative, you can wet the paws of your dog and then ask him to stand so you can measure the distance between his legs.

As a guide, it is best to choose equipment that it is just big enough for your dog so that he does not fidget and move around too much.

Quality

Fitness training equipment is available from specialist outlets and this ensures it is of high quality and is safe to use. You can find some similar items which are less expensive, but you need to proceed with extreme caution. It may look the same, but the materials are often poor in quality and the surface cannot be guaranteed as non-slip. If you are using equipment with a textured surface, or with bumps, you will find big variations which some more sensitive dogs will find off-putting. Bear in mind, a negative experience at an early stage in training can have long-term adverse results.

For an overview of all the equipment we have used for the exercises in this book, see page 199.

This dog is supporting himself on his knees, which is not desirable.

The trainer's eye

You may have the best theoretical knowledge when it comes to fitness training, but this is completely useless if you, as the trainer, do not develop an eye for what looks right, and for detecting potential problems. Ideally, you would train in front of a mirror so you can look at the dog's posture and correct where necessary. If this is not possible, you could enlist a helper to take photos, which will give you the opportunity to review your training and avoid potential pitfalls.

The dog is forced to compress himself, resulting in a hollow back. This could be avoided by adjusting the distance between the dog's front and back legs.

As mentioned previously, check the equipment and make sure you always work on a non-slippery surface. You may find it helpful to buy an exercise mat from a sports retailer and use this as your working base.

When you are training, observe your dog closely at all times. Make sure his back is straight, his legs are parallel, and he is working calmly. If an exercise is to be executed on an unsteady or wobbly base, the dog should first do it on a low, firm platform before

This Jack Russell cannot keep the tension in his body and slumps down on his bottom.

progressing to the greater challenge of a wobble cushion. Again, the dog should master this before further increasing the level of difficulty.

Some dogs are not, naturally, in tune with their bodies. If your dog is of this type, allow plenty of time for the set-up, and do not introduce the challenge of a wobbly surface too soon. Signs that the dog is struggling to balance include fidgeting, panting, jittery compensatory movements and muscle tremors, as well as attempting to increase the area of the body that is in touch with the equipment. Sometimes you will see the tail rowing in an attempt to stablise the body. If you see any of these signs, you should step in and offer assistance. It may be that you need to stabilise the equipment or support the dog. But, most importantly, you need to reduce the challenge as the dog may well be overwhelmed.

Let's take training the dog to stand on a Fitbone as an example:

Step 1: Place a board on top of two Fitbones, so the dog learns to stand on a wobbly base.

Step 2: Remove the board and place two Fitbones laterally. Encourage the dog stand with his front feet on one Fitbone, and his back feet on the other.

Step 3: Keep the first Fitbone in position for the front feet and place the other Fitbone lengthways so the dog has to place his back feet a little closer to each other.

Step 4: You are now ready for the dog to balance on a single Fitbone – but support him as much as needed.

Remember, that there is a thin line between encouraging or challenging your dog, and overwhelming him.

Warm-Up and Cool-Down

How to warm up correctly

Every kind of sport needs to start with a warm-up programme. A mental and physical warm-up prepares the dog for training, prevents injuries and improves the physical performance as a result of improved neuromuscular communication. A well-executed warm-up is the starting point for every successful training session.

In addition, a warm-up can also be seen as a team building exercise which will also increase your dog's motivation for training.

What happens during a warm-up?

- Circulation – increase of body core temperature and muscle temperature.
- Cardiovascular system – the heart beats faster and pumps more blood through the body so that the muscles are supplied with more nutrients and oxygen.
- Respiration – breathing intensifies and provides a better oxygen supply.
- Neural system – the central nervous system is activated, ensuring faster reactions, which improves response speed, and also reduces the risk of injury.
- Joints and ligaments – production of synovial fluid improves the elasticity of cartilage, thereby improving tolerance to strain. Due to the increase of body temperature, ligaments and tendons also become flexible.
- Muscles – are better supplied with blood and therefore become more efficient and less prone to injury.
- Brain – improves concentration and cognitive ability which, again, enhances response speed.

We distinguish between a general warm-up to increase body temperature, cardiac activity and breathing function to get the body to the ideal performance level (e.g. warming up by moving at different speeds) and a sport-specific warm-up, where we warm up certain parts of the body that will be put under particular strain by the forthcoming activity (e.g. co-ordination and speed strength in agility).

Length of warm-up: approx. 10 minutes

The length for a warm-up programme cannot be stipulated because it depends on the ambient temperature, the dog's performance level, and his age.

If you take your dog training early in the morning, before exercise, you will need to devote more time to a warm-up than if he had already been for a walk. If you have an older dog, and it's very warm outside, you need to make sure that you don't over-tire him during the course of the warm-up.

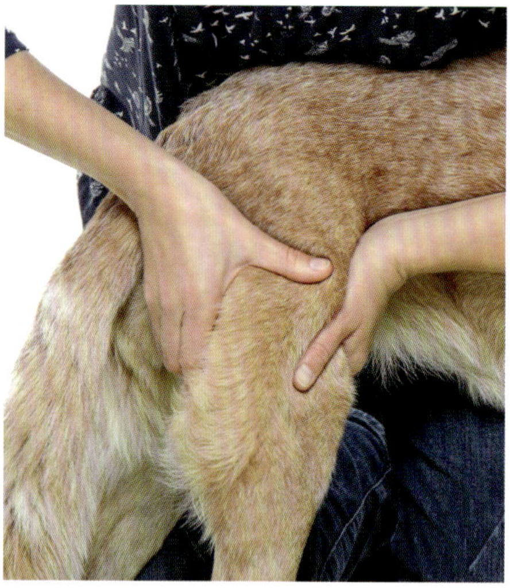

Massage the major sets of muscles.

If your dog does agility, it is important to move the toes as part of the warm-up.

If the dog gets into a heeling position immediately, it helps if the handler walks backwards.

Moving the dog's head to his flank promotes sideways mobility of the spine, and warms up the side muscles of the back.

In general, muscles should not be fatigued before training, as this poses a higher risk of injury. Training would also be ineffective as the muscles would not respond as required.

The four-step programme for warm-up

1. Walking

Attach a long lead (1-2 m/6-12ft) and allow the dog to toilet. Walk very slowly at first, then increase your pace a little, walking the dog on both your right and left side, before increasing your pace to a trot.

2. Massaging muscles

Massage the major sets of muscles: neck, back, upper arm, shoulder and thighs. Use a firm pressure and work at speed. You will have to be a little more gentle when you massage the feet. Work swiftly, as you do not want the massage to have a relaxing effect before you work with the dog.

3. Active mobilisation of the spine

Mobilise the cervical spine by moving the dog's head to the right, and left, shoulder a few times, and then move his head to each flank. Next, move his head up and down, and underneath his belly. It is easier if you do these exercises by luring with a treat or using a hand target.

Luring the dog's head underneath his belly mobilises the small facet joints of the spine and activates the long musculus erector spinae.

4. On the move

Sit-down-stand.

Crawling.

Walking backwards.

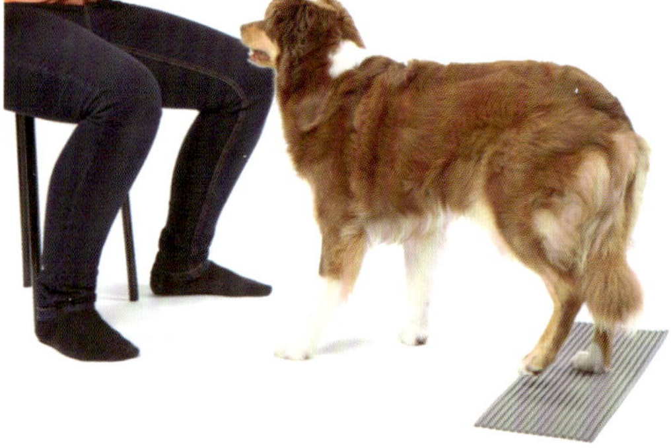

Taking lateral steps.

Moving in a figure of eight around the legs.

High-five or shake paws at different heights.

COOL-DOWN

The function of a cool-down is to return the body to its normal temperature and to remove waste products. It provides the opportunity to address structures that have been put under strain in the form of passive stretching. It is important to note that an off-lead walk with a dog buddy after a workout is not a cool-down; it is more like a second round in the ring.

Stretching for relaxation

The key factor for all stretching exercises is serenity. You need to find a quiet spot with minimal distractions, so the dog can relax both physically and mentally.

All stretching exercises, following activity, should last for at least 10 seconds so that the receptors in the muscles and tendons are activated and, therefore, relax the muscle. If the stretch is carried out too quickly, it leads to an increase of muscle tone, which makes it into a warm-up exercise.

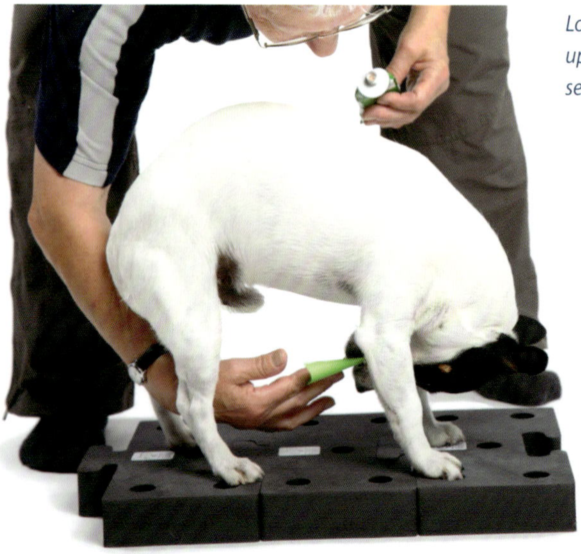

Looking under the belly is an active exercise that helps to warm-up the back muscles. The dog should stay in position for 10 seconds.

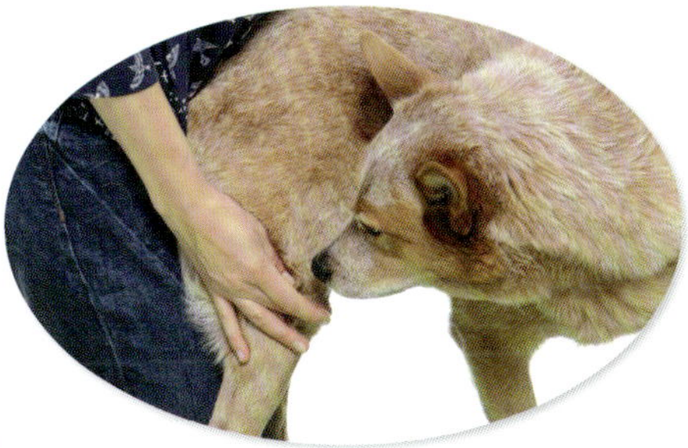

With the dog standing, use a target or a treat to move his head towards his shoulder. Hold for 10 seconds.

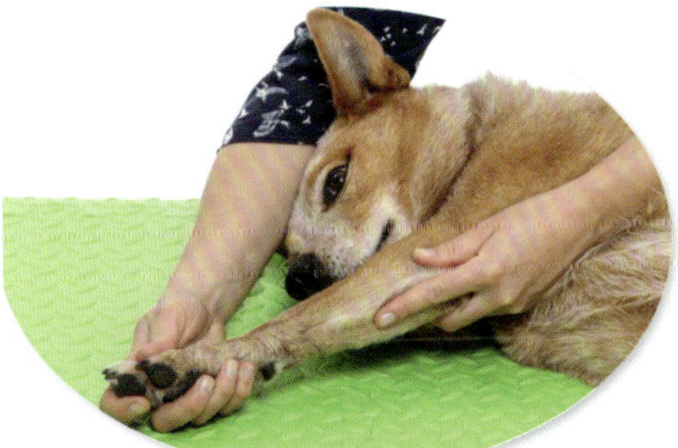

With the dog standing, use a target or treat to move his head sideways towards his knee. Hold for 10 seconds.

With the dog lying flat, move his front leg forwards in a stretched position. Hold for 10 seconds.

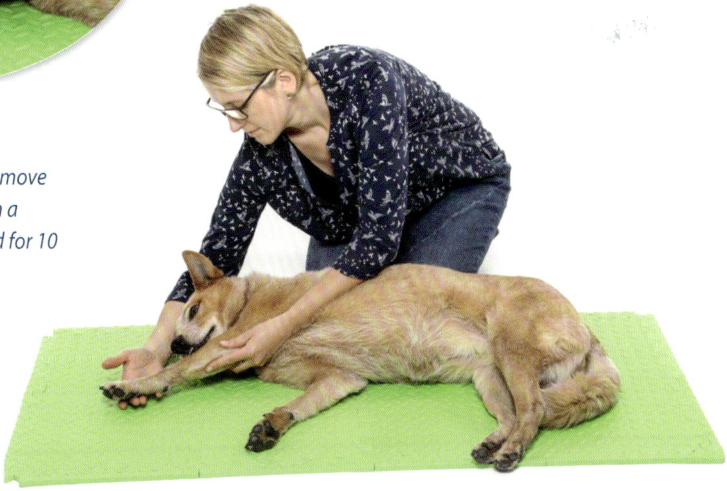

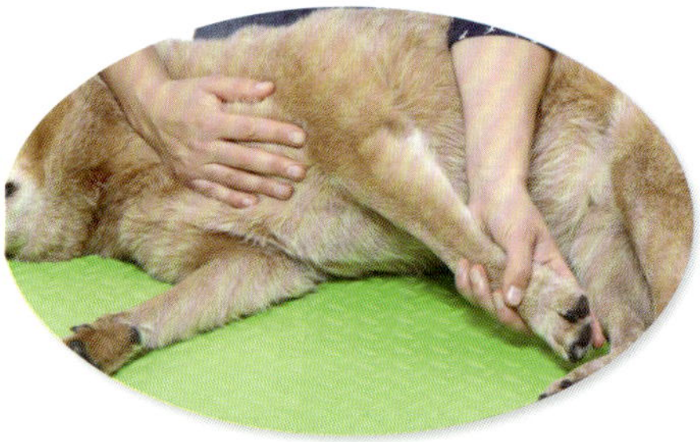

With the dog lying flat, move his front leg backwards in a stretched position. Hold for 10 seconds.

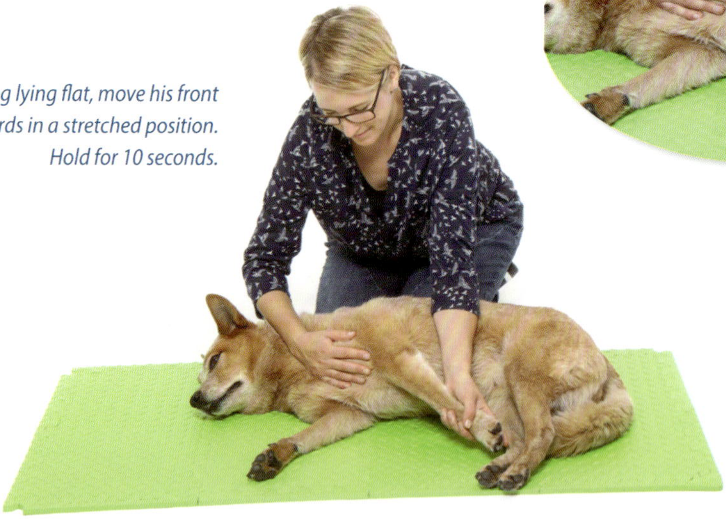

With the dog lying flat, move his hind leg backwards in a stretched position. Hold for 10 seconds.

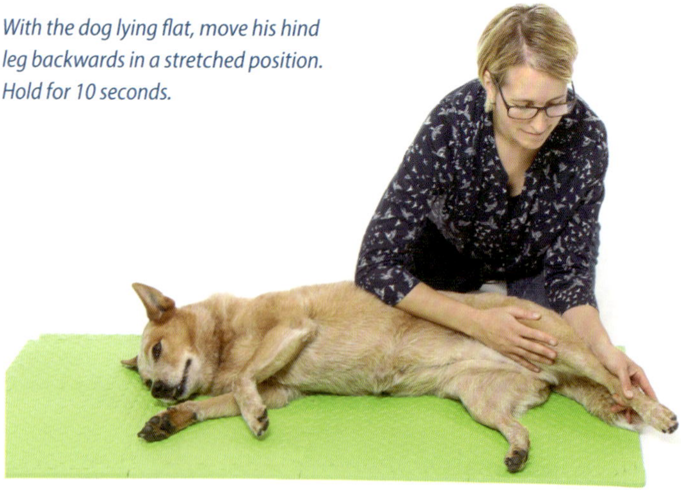

With the dog lying flat, the front leg is abducted (moved from the midline of the body) sideways in a stretched position. Hold for 10 seconds.

With the dog lying flat, move his hind leg sideways in a stretched position. Hold for 10 seconds.

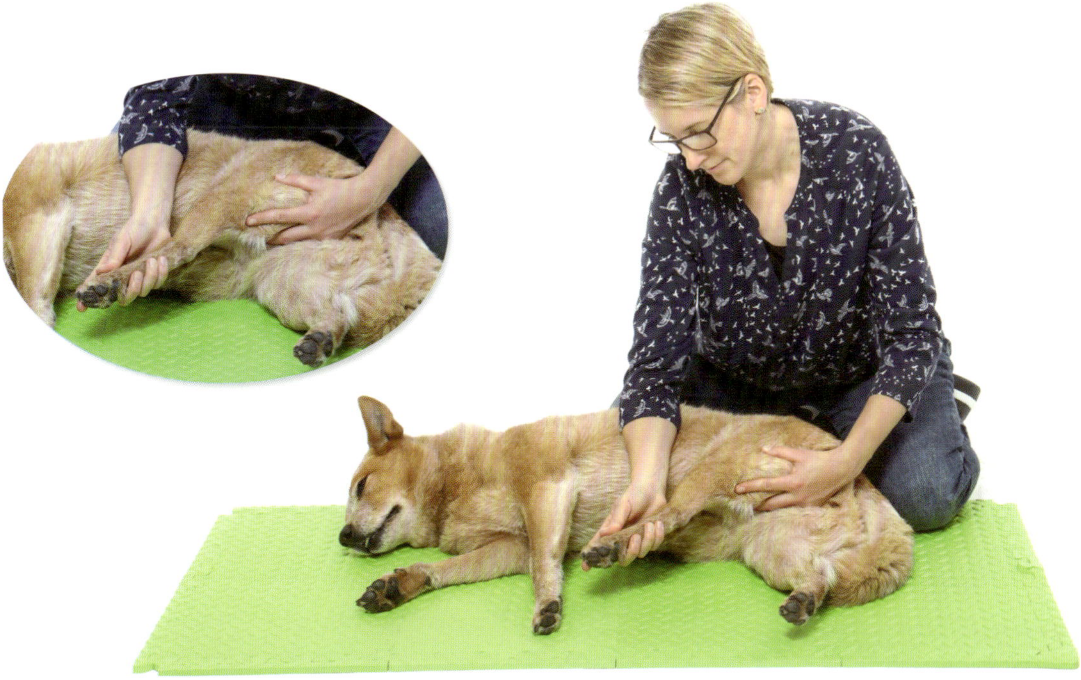

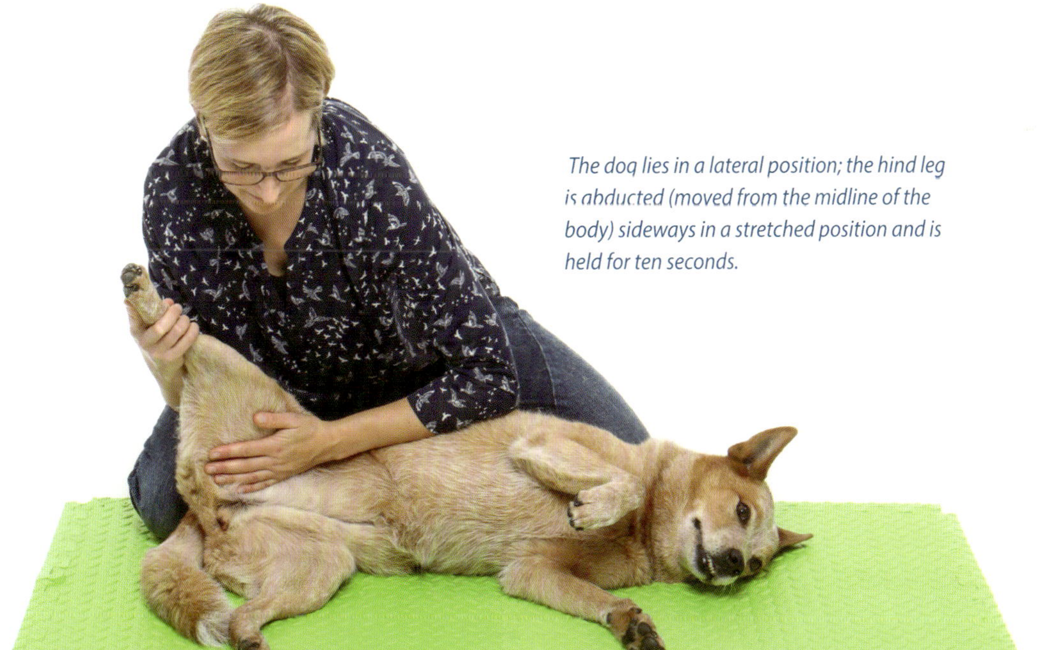

The dog lies in a lateral position; the hind leg is abducted (moved from the midline of the body) sideways in a stretched position and is held for ten seconds.

Active Exercises

We have categorised active exercises in relation to the specific areas of the body we are working on. We first provide an overview and then focus on the exercises in detail. We start at the front of the dog and work progressively to the back, as follows:

- Neck muscles and shoulder
- Fore limb
- Front thigh muscles
- Outer thigh muscles and gluteal muscles
- Back muscles part one – relaxation
- Back muscles part two – activation
- Pelvic floor
- Body awareness/coordination of the hind limbs

For each part of the body, there is a selection of exercises, and these are sub-divided as 'basic', 'advanced' and 'professional', labelling the degree of difficulty that is involved. Of course, this may vary, depending on the individual dog, so it is only intended as a rough indicator. However, the exercises are designed to progress from 'basic' to 'professional', so it makes sense to train them in this order, unless your dog is already advanced.

Orientation Buttons

In order to find the right exercises for your dog, at first glance, we have used orientation buttons above the exercises. This is intended to show you the focus of each exercise. The orientation buttons refer to the exercise itself, so it is, therefore, possible that there will be more then one benefit. For example, an exercise may strengthen the pelvic floor and improve coordination.

Buttons are categorised as follows:

STRENGTH

Strength

For these exercises, the dog has to work against a resistance, or he has to hold a position without changing the length of the muscle (see Isometry, page 6). They are designed to improve muscle development, as long as the dog is not experiencing muscular tension. Bear in mind, these exercises are exhausting so make sure the dog is rested before beginning work.

RELAX

Relaxation

These exercises are designed for muscle relaxation and, as with all stretching exercises, they need to be held for at least 10 seconds. They can be scheduled for the end of a training session; do not ask for them before the strength or co-ordination training.

CO-ORD

Co-ordination

Exercises for co-ordination and proprioception promote the body's ability to respond, and adjust, to complex movement sequences. They also improve body awareness, balance and dexterity. If a young dog was confined to walking on pavements and flat surfaces for the first year of his life, he would find a hike over steep and rough terrain much more challenging than a dog who had was accustomed to co-ordination exercises. These exercises are, to some extent, important for all dogs, regardless of whether they are involved in a sporting activity.

FLEX

Flexibility

The exercises in this section are designed to improve mobility and extension. Cavaletti training, for example, will help to develop mobility skills.

TACTILE

Tactile/Sensory

In order to address neural stimuli, you can use textured massage balls to improve body and tactile awareness. These exercises are especially important for oldies and dogs with neurological problems, and also for dogs with poor body awareness.

II. EXERCISES FOR THE NECK AND SHOULDER MUSCLES

This is an overview of the exercises that are outlined in detail on the following pages.

Neck Muscles and shoulder area		
Category:		***Cervical Spine***
Aim:		*Strengthening or relaxation, mobilisation and co-ordination*
Recommended:		*For dogs with limited mobility of the neck muscles, and to stabilise strain due to sporting activity*
		Unfortunately, neck muscles and muscles of fore limbs, with shoulder blade and torso, cannot be separated clearly, so there will be overlaps in these categories
		Hold position for 10 seconds or longer = relaxation *Swift execution = muscle development (also for warm-ups)*
Muscle function:		• *All movable muscles of the head.* • *Musculus brachiocephalicus (head-arm-muscle) – forward movement of fore limbs, muscle for locking in position, sideways bending of the neck, pulls head and neck downwards* • *Musculus splenius capatis (long head muscle) – extensor and for lifting, muscle for sideways bending of neck and head* • *Musculus longus colli (neck-flexor) – flexor for cervical spine* • *Musculi scalene (middle and back muscle for holding the ribs) – flexor, sideways bending and muscle to help with breathing* • *Musculus longissimus (long back-, neck- and head muscle) – for lifting, extensor, muscle for locking in position and sideways bending* • *Musculus spinalis et semispinalis thoracis et cervicis – for lifting, muscle for locking in position, extensor spine and muscle to straighten torso* • *Musculus transversospinalis – for lifting, muscle for locking in position, extensor and sideways bending of head and neck.* • *Musculi multifidi (multi branched or split muscles) –muscle for locking in position, turning and lifting* • *Musculus iliocostalis (stabilises and straightens the spine) does not concern the neck as much, but active while lateral bending.* ▸▸ *Pectoral girdle (see fore limbs)* ▸▸ *Core muscles (see fore limbs)* ▸▸ *Side abdominal muscles for sideways bending*

Neck-shoulder muscles ①	FLEX +++ RELAX +++	***1.) Nose to chest (Basic)*** *The dog is in a sitting position and brings his nose down to the sternum. 3-4 repetitions; for stretching: hold still for at least 10 seconds* ▸▸ *Stretches or trains extensor of cervical spine and head lifter*
	FLEX +++ RELAX +++ CO-ORD ++	***2.) Looking underneath the belly (Advanced)*** *Dog is in a standing position and, with the help of a target, the nose is tucked under the belly* ▸▸ *Stretches or trains extensor of cervical spine and head lifting, as well as the extensor of thoracic spine. Also activates the abdominals* *3-4 repetitions; for stretching: hold still for at least 10 seconds – however, abs are also activated here*
Neck-shoulder muscles ②	FLEX +++ RELAX ++	***1.) Stretched sphinx (Basic)*** *Dog is in a sphinx-like position and, with the help of a target, stretches forward* ▸▸ *Stretches the muscle for bending the neck, relaxes the head lifting muscle, extensor of the cervical spine and relaxes the thoracic muscle loop* *Hold stretch for at least 10 seconds to relax muscles*
	FLEX +++ RELAX +++	***2.) Slanted sphinx (Basic)*** *Same exercise with focus sideways left and right, 3-4 repetitions* ▸▸ *In addition, exterior shoulder muscles and muscle for tilting sideways (of the opposing side) are stretched.* *Hold stretch for at least 10 seconds to relax muscles*
	FLEX +++ RELAX +++	***3.) Wobbly sphinx (Basic)*** *Position the dog on a wobbly surface, such as a peanut ball, 3-4 repetitions* ▸▸ *This also trains co-ordination*
Neck-shoulder muscles ③	FLEX +++ RELAX ++ CO-ORD ++ STRENGTH ++	***1.) Elevated stand (Basic)*** *Fore limbs are elevated, and the dog is fed upwards, at a slanted angle (straight backline), hold position for a few seconds* ▸▸ *Front is 'opened up', stretching head-arm-muscle, relaxation of neck extensor as well as stretching abdominals and activating muscles of the hind legs*

Neck-shoulder muscles ③	RELAX + CO-ORD +++ STRENGTH ++	**2.) Mountain goat (Advanced)** Position the dog on a steady platform slanted upwards (straight back line) ▸▸ Stretching of head-arm-muscle, neck and pectoral muscle, activation of abdominals and back muscles
	CO-ORD +++ STRENGTH ++	**3.) Mountain goat/wobbly (Professional)** Same exercise but use a wobbly surface, such as a donut ▸▸ Increased strengthening of head-arm-muscle, neck, pectoral and back muscle
Neck-shoulder muscles ④	FLEX +++	**1.) Head to shoulder (Basic)** From a standing position, the dog's head is lured sideways towards the shoulder blade ▸▸ Stretching of muscle bending sideways, mobilisation of cervical spine 3-4 repetitions; for stretching hold for at least 10 seconds
	FLEX +++ RELAX +++ CO-ORD ++	**2.) Head to knee (Basic)** From a standing position, the dog's head is lured sideways towards the knee, parallel to the shoulder blade ▸▸ Stretching neck muscle responsible for sideways bending, side muscle of cervical spine of pectoral muscles and lumbar spine, side abdominals, mobilisation vertebra and rip joints 3-4 repetitions; for stretching hold for at least 10 seconds
	FLEX +++ CO-ORD +++ STRENGTH +	**3.) Head to shoulder or knee/ unsteady (Advanced)** Same exercise but position the dog on a slightly unsteady surface. 3-4 repetitions ▸▸ Additionally trains co-ordination
	FLEX +++ CO-ORD +++ STRENGTH ++	**4.) Head to shoulder or knee/ wobbly (Professional)** Same exercise, but on a peanut ball, 3-4 repetitions ▸▸ Additionally trains co-ordination

Neck-shoulder muscles (5)	CO-ORD + STRENGTH +	*1.) Hold (Basic)* *Dog pushes his chin into your hand (pre-exercise 'hold')* *▸▸ Strengthening muscle for bending the neck and relaxing the neck muscles (intensity depends on how hard the dog pushes)*
	CO-ORD ++ STRENGTH +	*2.) Balancing an object on the neck (Advanced)* *Dog balances an object on his neck while doing 'hold', can be used as distraction for dogs that stay in protective postures* *▸▸ Additional focus on object and on balance*
Neck-shoulder muscles (6)	CO-ORD ++ STRENGTH ++	*1.) Nodding with dumbbell (Basic)* *Dog learns to look up and down while holding a dumbbell* *▸▸ Exercise for strengthening of neck muscles*
	CO-ORD ++ STRENGTH ++ FLEX ++	*2.) Nodding with dumbbell/wobbly (Advanced)* *Dog performs the basic exercise on a peanut ball or something comparable.* *▸▸ Additional focus on dumbbell and balance*

FLEX +++

RELAX +++

Exercise Band 1

1. Nose to chest (Basic)

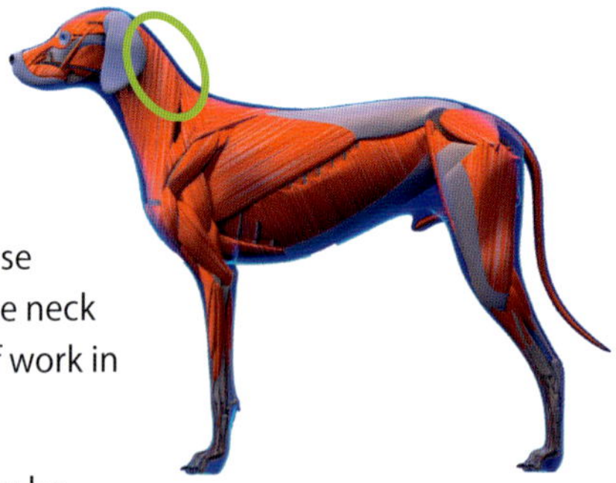

The dog is in a sitting position and brings his nose down to the sternum. This exercise addresses the neck muscles, compensating for dogs who do a lot of work in the heelwork position.

1. Ask the dog to sit on a small, raised platform; he should be able to sit comfortably, but the platform should have sufficient length if he were to stand. Reward the dog a couple of times for a straight sit, and then position yourself next to him.

2. Depending on the dog, work with a chin target ("hold"), nose target or a high value treat (e.g. liver paste). Refresh your hand or chin target in the sitting position, and reward a few times. If you are working with a treat, you can skip this step as you will be luring the movement. However, the act of luring can cause the dog to fidget, and the neck muscles may not fully relax.

3. Step by step, work in a quadrant (quarter of a circle) starting at the tip of the dog's nose and working towards the sternum. Do not work too fast and really picture

the quadrant in your head. Work in steps of 10 degrees at a time, rewarding each step, and then return to the starting position so that you can begin anew.

4. If the dog is only following your hand up to a certain point, it could mean that he is experiencing tension somewhere. Please take note of this behaviour and give him a break. If the dog keeps showing this behaviour in subsequent sessions, try to increase motivation with some especially tasty treats. Sometimes the dog is simply being lazy and needs encouragement, but it could also be caused by restriction of movement, which will need further investigation.

5. If you are training this exercise with a chin target, your hand's end position is on the dog's sternum; if you are working with a nose target, the dog should be aiming for the side of your hand.

6. Now you can work on duration by delaying the click and feeding in position.

7. If your dog is doing the exercise reliably, you can introduce a verbal clue and reduce the help with your hand, a little at a time.

8. Make sure the dog sits correctly throughout the exercise, otherwise you will not achieve the full effect.

The dog's nose is lured to the sternum with the help of a tube filled with liver paste, or you can work with a target. Make sure the head stays straight.

FLEX +++

RELAX +++

CO-ORD ++

Exercise Band 1

2. Looking underneath the belly (Advanced)

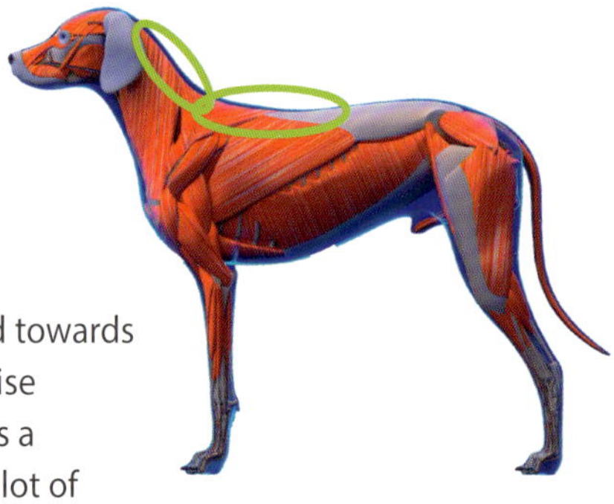

From a standing position, the dog's head is lured towards his sternum, with the help of a target. This exercise addresses the neck muscles and the abs, too. It is a good compensatory exercise for dogs who do a lot of heelwork. Position your dog on a platform to make sure that he remains in a standing position throughout the exercise.

1. Practise the nose target with the help of a Post-it note.
2. Offer the Post-it to the dog as a target that he must only touch with the nose. Click and reward.
3. Once you have established the nose touch, gradually lower the Post-it towards the ground. Click and reward each step.
4. Now, position the Post-it a short distance from the dog's front legs. He should be able to reach it easily with his nose, without moving from the stand. This step is very important, so train it well.

The aim is for the dog to bend his neck to the extent that he can touch a target, held underneath his belly. As an incentive, you can put some liver paste on the target. Make sure the dog's head is in a straight line and not twisted sideways.

5. Now, position the Post-it directly underneath your dog by moving your hand through his hind legs. If he checks what is going on underneath his belly, click his interest. It is very important to feed the dog in this unusual position.

6. Position the Post-it underneath the dog's belly, just in front of his hind legs. If he fails to show interest in it, you can point to it, or even place a treat on it. Click and reward a glance towards it. He does not have to (and may be unable to) physically touch it. Sitting down or turning around is basically impossible due to the set-up of the exercise, as your dog is positioned on a platform.

Tip: If the dog is uncomfortable with you moving your arm underneath his belly, you can use a target stick for the set-up of the exercise.

7. Work towards the dog achieving an arch of the back by progressing from a glance through the legs, to holding his head level with his front legs.

Tip: As an alternative, you can position each front leg on a platform, which will enhance understanding.

8. Introduce a verbal cue and gradually remove the target from your set-up.

Exercise Band 2

1. Stretched sphinx (Basic)

+++ FLEX

++ RELAX

The dog starts in a sphinx position and stretches forward. This exercise addresses the upper back muscles and the lower neck muscles. It can be taught using a target or by luring. Here are the steps for working with a target:

1. Practise your hand target (or use a target stick) several times and reward the dog.
2. Ask him to lie down in the correct sphinx position.
3. Offer the target right in front of the dog's nose so that he does not have to move forward, but only stretches his head forward.
4. Offer the hand target (or target stick) a little further away, gradually increasing the distance until the dog has to stretch fully. Mark and reward at every stage. If he crawls to reach it, or even gets up, the target is too far away and you will need to go back a step.
5. This exercise is governed by tension and relaxation, which means that you should give your dog a treat when he is in the relaxed, starting position, and, again, when he has touched the target. By nudging the target, the dog will do a kind of a teetering motion, which you can best see at the croup.

The dog should not tilt sideways while in the sphinx position.

FLEX +++

RELAX +++

Exercise Band 2

2. Slanted sphinx (Basic)

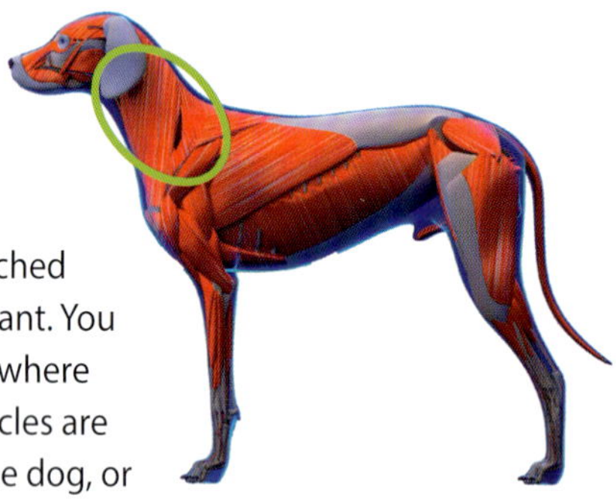

The dog is in the same position as with the stretched sphinx but, this time, he is lured forwards on a slant. You need to make sure that the head is in a position where the spine can still be straight. The side neck muscles are addressed in this exercise. Again, you can lure the dog, or you can use a target, as follows:

1. Ask the dog to lie down. Throughout the exercise you should make sure his hind legs remain in the sphinx position and do not tilt sideways. If it does not bother your dog, you can put one hand on his croup, or right next to his paws, as a reminder.

2. Practise your hand target, or use a target stick.

3. Position the target to one side of the dog and encourage him to stretch towards it. Do not ask too much of him or he will get up or tilt sideways.

4. Make sure to position the dog in a way that the spine does not 'deviate' but, rather, bends sideways while staying in a line. The dog should not move his feet.

5. The ideal stretch is achieved just before the dog needs to move his hind legs to follow the target. However, they should always remain in the sphinx position.

6. Once the dog has achieved a good stretch on one side, train the other. For activation, repeat the exercise three to four times. You can take turns on the sides. For relaxation, the dog should hold the position for at least 10 seconds.

For this exercise, lure the dog's head forward and sideways.

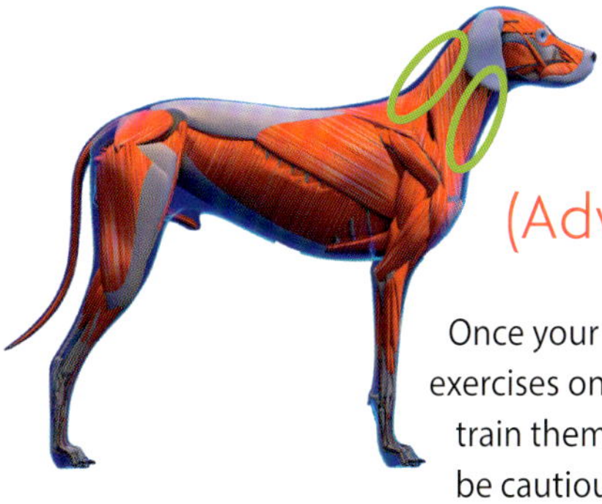

Exercise Band 2

3. Wobbly sphinx (Advanced)

+++ FLEX

+++ RELAX

Once your dog is comfortable with both basic sphinx exercises on a stable surface, you can train them on a wobbly surface. Please be cautious and work slowly on this transition. We recommend you start with a soft surface, such as a trampoline, before progressing to a very wobbly surface, such as a peanut ball.

The Peanut ball should be secured for this exercise.

Equipment: FitPaws Peanut ball (sensory).

FLEX +++

RELAX ++

CO-ORD ++

STRENGTH ++

Exercise Band 3

1. Elevated stand (Basic)

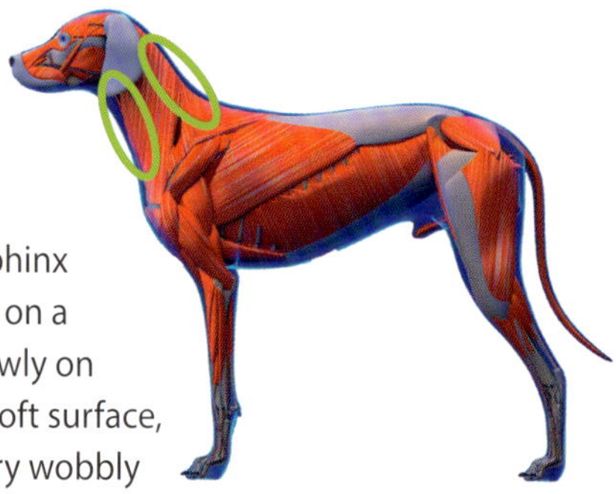

Once your dog is comfortable with both basic sphinx exercises on a stable surface, you can train them on a wobbly surface. Please be cautious and work slowly on this transition. We recommend you start with a soft surface, such as a trampoline, before progressing to a very wobbly surface, such as a peanut ball.

1. Use a stool, or something similar, that will really elevate the dog when he places his front feet on it.
2. Initially click and reward the dog for showing even the slightest interest in the stool (e.g. sniffing).
3. If the dog does not progress to putting his front feet on the stool, help him by offering a target.
4. Click and reward each time the dog gets into position until he reliably offers the behaviour.
5. Now you will need your hand target. Offer it above the dog's head, and to the side. The dog should be able to touch it in such a way that you could draw a line from the tip of his nose to his tail. This is what you are looking for in an ideal stretch.
6. You can, at this stage, introduce a verbal cue.

Make sure the dog stands parallel and has a straight backline.

Equipment: IKEA stool (tall).

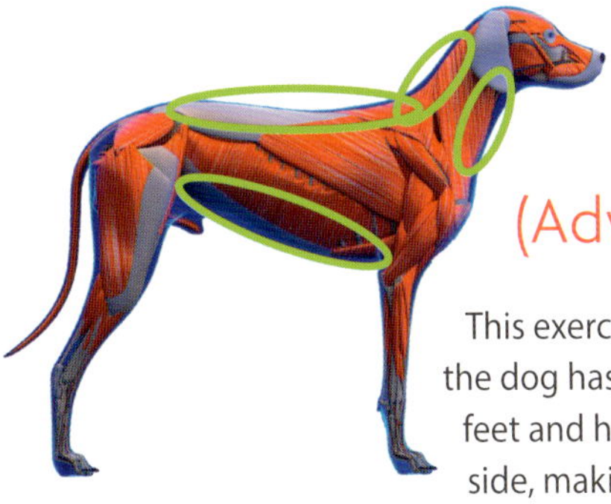

Exercise Band 3

2. Mountain goat (Advanced)

\+ RELAX

+++ CO-ORD

++ STRENGTH

This exercise is executed on a stool, or a platform, so that the dog has to stand with a narrow gap between his front feet and his back feet. Again, feed upwards, and to the side, making sure the backline is straight. This exercise stretches the head-forelimb muscle, activates the back muscles and abs, and the femoral muscles.

1. Similar to *elevated stand* (see page 56), this exercise is mainly concerned with stretching the head-forelimb muscle. The platform/stool should have just enough space to accommodate all four paws.
2. Present the stool and make it interesting before you even set it down. Click and reward the dog's first interest, gradually progressing until he climbs on it. If necessary, help him with a hand target.
3. The most difficult part of this exercise is to get the dog to put all four paws on the stool. Give him the time he needs to feel comfortable and safe, and can fully balance when he is standing on it.

For this exercise, the dog has to narrow the gap between his front and back feet.

Equipment: IKEA stool (tall).

4. When the dog is in position, feed him, or offer a target, upwards and to the side. Be guided by the dog's spine; he should not have to bend his neck to reach your hand target.

5. Once the dog has to stretch so far that he is almost struggling to reach your hand target, you have found the right position. This is a very intense workout, so allow breaks between asking for this exercise, regardless of whether the dog is still learning or already knows it well.

CO-ORD +++

STRENGTH ++

Exercise Band 3

3. Mountain goat/wobbly (Professional)

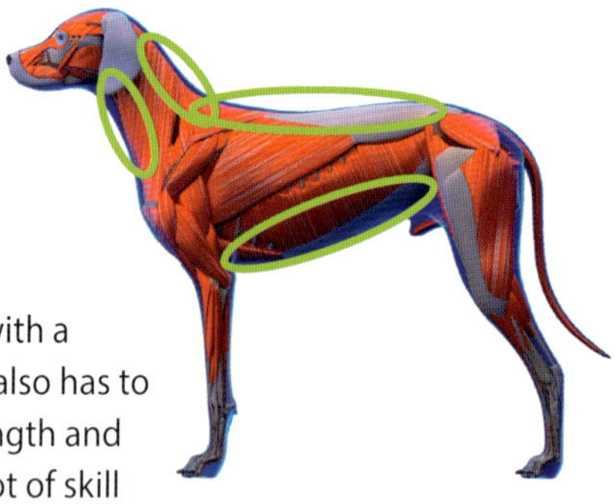

For this exercise, the dog not only has to stand with a narrow gap between his front and back feet, he also has to balance. We are working on strength and co-ordination, and it requires a lot of skill to hold position.

Once the dog is very comfortable with performing *mountain goat* on a stable surface (see page 57), you can ask him to do it on a wobbly surface. Please note that this exercise is very intense and should be reserved for dogs who are fully fit.

To teach wobbly mountain goat, initially work on a bigger surface area, and progress to a smaller area as the dog gains in confidence.

Equipment: Stabilised Therapy Gyroscope.

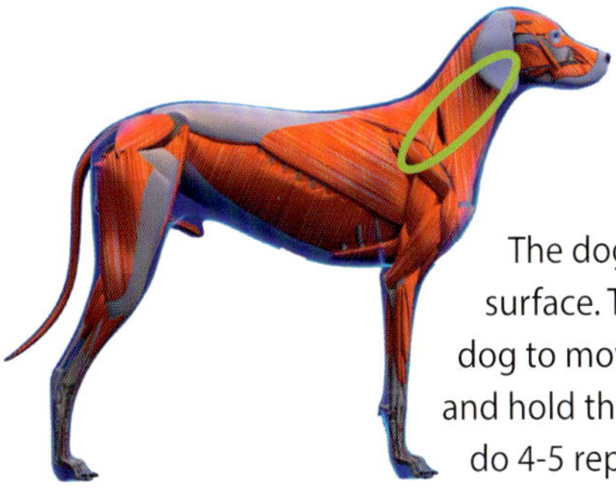

Exercise Band 4

+++ FLEX

1. Head to shoulder (Basic)

The dog is positioned in a stand on a flat, non-slip surface. Then, using a treat or a hand target, encourage the dog to move his head sideways, towards the shoulder blade, and hold the position. If you use this exercise for a warm-up, do 4-5 repetitions swiftly. If you use it for active stretching, hold the position for at least 10 seconds.

1. Practise a hand target.
2. For this exercise the dog needs to stand parallel, so ask him to step on to a platform that has limited space, on either side, which will prevent the front legs shifting. Click and reward several times for adopting the correct start position.
3. Offer a hand target to your dog and reward him for nudging it.
4. Step-by-step, move the hand target from the tip of your dog's nose towards his shoulder blade. Reward each single step.
5. Once you have reached the point at which the dog is bending towards his shoulder, train the other side.
6. For activation, repeat each side three to four times; for stretching, the dog should hold the position for at least 10 seconds.

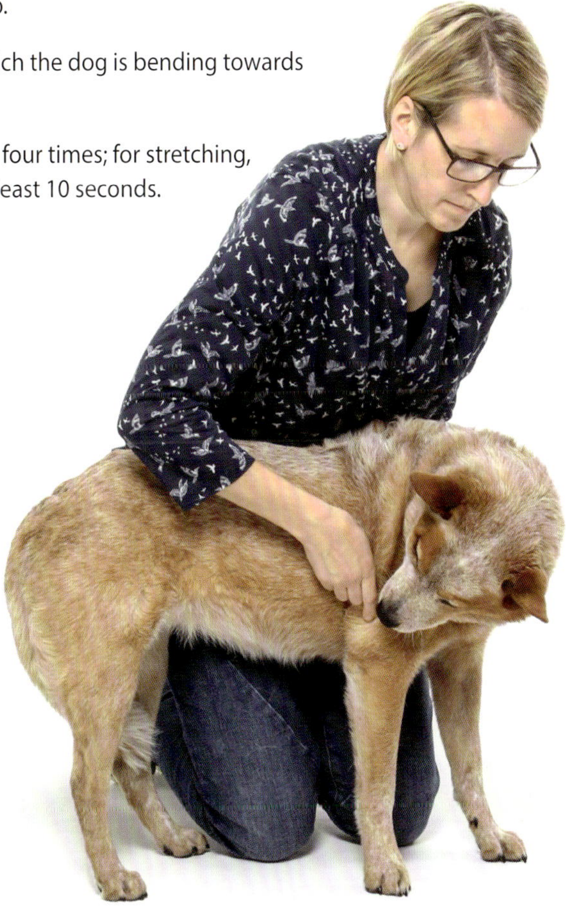

You can support the dog on his opposite side.

FLEX +++

RELAX +++

CO-ORD ++

Exercise Band 4

2. Head to knee (Basic)

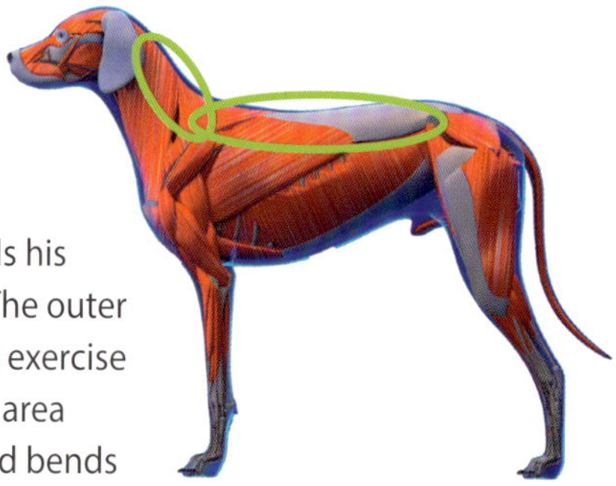

The dog is encouraged to move his head towards his knee, with the help of a treat, or using a target. The outer shoulder area is additionally addressed with this exercise and it also stretches the side, back and shoulder area (M.latissimus dorsi, M.trapecius). If the dog's head bends towards his left knee, you are training the right side of his neck/shoulder muscles, and vice versa when he bends to the right.

As with *head to shoulder* (page 59), make use of a platform to give your dog a visual restriction, even though, this time, he is allowed to move his forehand. Position yourself behind the platform.

1. Practise the hand target. If your dog already knows *head to shoulder*, ask him to do it several times.
2. Again, work gradually, rewarding each step. Take it slowly as the dog will not be able to reach his knee within the first session. It is a process that requires increased mobility, along with a full understanding of the exercise.
3. Once the dog is comfortable reaching his knee on one side, you can move on to the other side.
4. For activation repeat three to four times in quick succession. You can also combine it with the *head to shoulder* exercise. If you want to stretch the muscles, your dog should hold position for at least 10 seconds each time.

Tip: If the dog tries to turn around completely, place a hand on his thigh to encourage him to stand still. As an alternative, you can work with an additional target for the hind legs.

You can help your dog to get in position with a treat or by using a target.

Exercise Band 4

3. Head to shoulder/ unsteady (Advanced)

+++ FLEX

+++ CO-ORD

+ STRENGTH

If your dog is comfortable performing *head to shoulder* (page 59), you can transfer it to a wobbly surface which will train co-ordination as well.

For advanced dogs, you can transfer the exercises to a slightly unsteady surface.

Equipment: AIREX Balance Beam.

FLEX +++

CO-ORD +++

STRENGTH ++

Exercise Band 4

4. Head to shoulder or knee/wobbly (Professional)

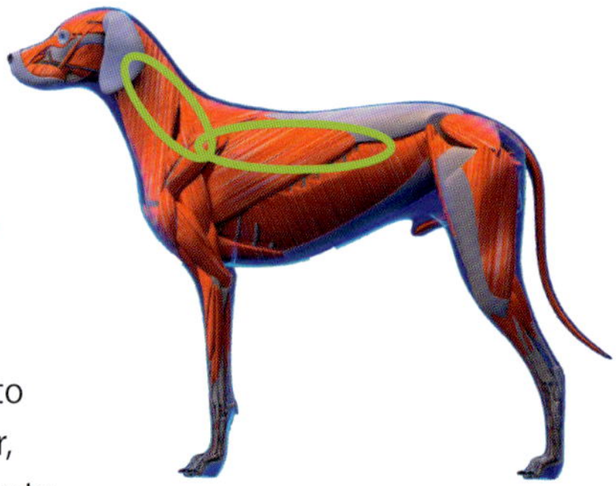

Both exercises – *head to shoulder* (page 59) and *head to knee* (page 60) – can also be transferred to a wobbly surface, such as a peanut ball. However, this is only suitable for dogs who have learnt how to balance. Performing this exercise in a lying position is easier to begin with but, later on, you can try it in a standing position.

Real 'professionals' find it easy to balance, even on a very wobbly surface.

Equipment: TOGU Dynair Extreme.

Exercise Band 5

1. Hold (Basic)

\+ CO-ORD

\+ STRENGTH

We have have given instructions for the *hold* among the preliminary exercises (see page 30), but is also essential as a neck and shoulder exercise.

You should be at eye level with your dog, and he will need to be in a stand. This exercise can be a little intimidating for more sensitive dogs, so be tactful when you are asking for it.

1. Show the dog that you have brought some top-quality treats for today's training, but do not give him a treat just yet.
2. Face the dog and cup your inside hand to provide a place for his muzzle, or his whole head (depending on his size). Hold a treat in your other hand.
3. Now place your cupped hand directly between the dog's muzzle and your other hand, which is holding the treat. This means the dog only has to move slightly towards the treat and he will, automatically, slide his muzzle into your cupped

Work on this exercise in easy stages so the dog learns to press down on your hand and hold the position.

hand. You have to vary the distance, and position of the treat, until you find the right position. Click and reward the moment the muzzle is in contact with your hand, making sure it is lying flat.

4. Practise a few times, clicking and rewarding each time the dog places his muzzle in your cupped hand.
5. You are now looking for the dog to push his muzzle into your hand, and to bear down so he is putting some weight into it. To achieve, this, repeat the previous step but, this time, move the hand holding the treat downwards. This will involve a slight delay in giving the treat. This manoeuvre requires some skill, so work on it in easy stages. The aim is for the dog to keep his muzzle in your hand, but to look downwards, which means he will press his muzzle into your hand. Click the moment he does this and give him a treat as he holds the position. Repeat several times until the dog understands the concept.
6. Progress by offering your cupped hand lower and lower. As an alternative, you can start moving your hand downwards when the dog has already fully placed his muzzle in it.
7. Try to lengthen the period of time in which the dog stays in position, pressing down on your hand. Most dogs find this aspect relatively easy.
8. Now introduce a verbal cue and fade the treat gradually by using it on a random basis.

CO-ORD ++

STRENGTH +

Exercise Band 5

2. Balancing an object on the neck (Advanced)

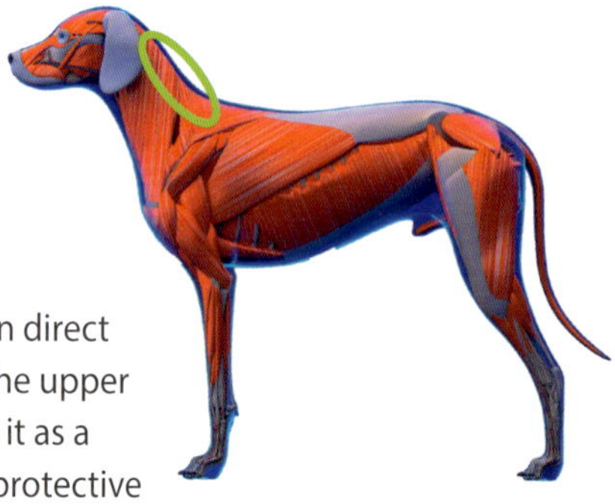

By balancing an object on the dog's neck, we can direct his focus on to this, which will allow us to train the upper and lower neck muscles. In addition, we can use it as a distraction from other 'problematic' areas, such protective postures. For example, the dog may have had an injury on one side which has completely healed, but his pain and/or muscle memory means he is reluctant to fully engage that side.

1. For this exercise, it is best to sit down on a chair and let the dog stand in front of you.
2. Offer a hand for *hold* – click and reward.

3. While the dog holds position, with his head in your hand, place a light object, such as a glove, on his neck. Click and reward if he co-operates. If the dog tries to see what you are doing, you can allow it. Simply try again.

4. Once the dog has accepted the object on its neck, you can stop holding it in position. Reward the dog for standing still.

5. Once the dog is balancing the object on his neck, without help, slowly remove the hand that is cupping his head. Proceed as before, and then remove your hand, finger by finger. Click and reward each step.

For this exercise,you can station your dog on an elevation.

Equipment: Togu Balanza Ballstep

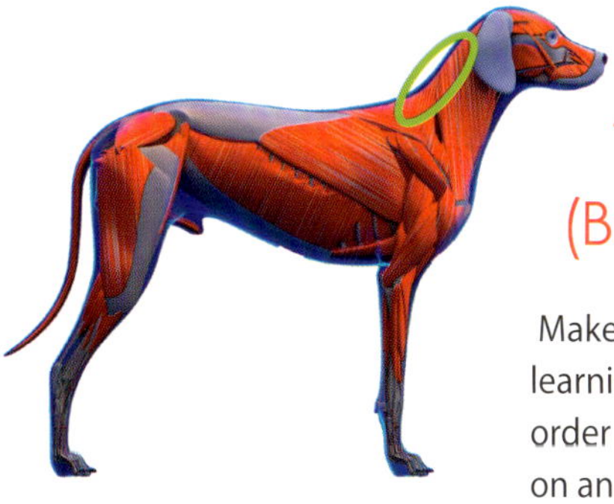

Exercise Band 6

1. Nodding with dumbbell (Basic)

++ CO-ORD

++ STRENGTH

Make sure the dog is in a standing position while learning this exercise so that he cannot lie down in order to reach the target. It can help to position him on an elevation, such as a dog stepper.

1. Prepare yourself by making sure your clicker, treats and Post-it notes are at hand.

2. Offer the dog a Post-it as a target for a nose touch – click and reward. Some dogs prefer working with their paws, so make sure you hold the target high enough to prevent this happening.

3. Once you have established the nose touch, gradually move the target towards the floor. Click and reward each step.

4. Now put the target on the floor, a little distance from the dog's front legs so he can reach it without moving.

5. Be consistent, and only click and reward the dog for touching the target with his nose. Hold the treat high enough so he has to stretch his head upwards to get it. This prevents him from lying down.

6. After eating the treat, the dog will look down forwards and touch the target again. Click this, and reward upwards again.
7. Introduce a verbal just before the dog restarts the exercise.
8. Proceed by gradually reducing the size of the target. Progressing too quickly may cause problems. If this happens, feed the dog on the target to rekindle his interest.
9. Now try the exercise without using a target. Click the dog at the lowest point of moving his head downwards. He will look up again, automatically, to receive his treat.
10. Once the dog fully understands this exercise you can progress to asking him to hold a dumbbell. Start from step 1, and work through each step, rewarding generously. Most dogs are quick to understand the correlation between the two exercises.

FLEX ++

CO-ORD ++

STRENGTH ++

Exercise Band 6

2. Nodding with dumbbell/wobbly (Advanced)

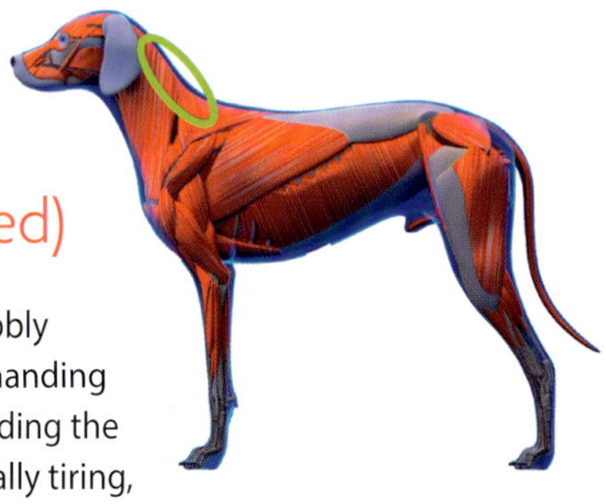

You can now transfer the basic exercise to a wobbly surface. However, be very cautious. This is a demanding exercise as the dog needs to concentrate on holding the dumbbell and keeping his balance. This is mentally tiring, and, because different forces are brought into play, it is also physically tiring. Therefore, this exercise should be trained for only a short amount of time.

Here, the dog is maintaining his balance while holding a dumbbell. This takes a lot of training, and you will need to proceed with caution

Equipment: TOGU Dynair Extreme

III. EXERCISES FOR THE FORE LIMBS

Forelimbs		
Category:		**Shoulder girdle / shoulder blade and forelimbs**
Training goal:		*Strengthening, co-ordination and improving range of movement*
Recommended:		*For dogs with under-developed shoulder muscles, for stabilisation during sporting activity and to improve step length*
Muscular function:		*• Musculus trapezius (trapezius muscle) – stabilisation and forward movement of the shoulder blade* *• Musculus brachiocephalicus (head-forelimb muscle) – forward movement of the limbs, sideways bend of neck and downwards bend of head* *• Musculus omotransversarius (shoulder-neck muscle) – forward movement of the limbs, muscle for sideways pulling of the head and neck* *• Musculus latissimus dorsi (broad back muscle)– retraction of the limbs* *• Mm. pectorales superficiales (pectoral muscles) –torso support, for bending sideways, forward and backward limb, extensor and muscle for locking shoulder joint, important when turning on the forehand. Lessens with age* *• Rhomboid muscle –stabilisation, lifts shoulder, forelimbs and neck, moves shoulder blade forward and locks in position* *• Musculus serratus ventralis (ventral toothed muscle) – stabiliser in the stance phase, muscle to help with breathing, suspension of the torso, total limb projection and retraction, sideways bending of the neck* *• Musculus supraspinatus (upper bone muscle) – stretches the shoulder* *• Musculus infraspinatus (lower bone muscle) –provides lateral stability to the shoulder joint, ensures the flexion of the shoulder and its slight external rotation* *• Deltoid muscle – stabilises the shoulder joint* *• Musculus biceps brachii (biceps brachii muscle) – extensor of the shoulder joint and flexor of the elbow joint* *• Musculus brachialis (upper arm muscle) – flexor of the elbow joint and rotates limb outwards* *• Musculus triceps brachii (three-headed upper arm muscle) – extensor of the elbow, flexion of shoulder joint, important muscle in the stance phase* *• Musculus tensor fasciae antebrachii (Tensor forearm fascia) – extensor of the elbow joint*

Forelimbs ①	STRENGTH +++ TACTILE +	*1.) Isometry with hand on shoulder (Basic)* ▸▸ *Strength exercise for the entire musculature of the fore limbs*
	STRENGTH +++ CO-ORD ++	*2.) Isometry with hand on shoulder/wobbly (Advanced)* ▸▸ *Strength exercise for the entire musculature of the fore limbs*
Forelimbs ②	CO-ORD + FLEX ++	*1.) Lateral cavaletti (Basic)* ▸▸ *Trains the muscles of the shoulder girdle, chest muscles and fore limbs*
	CO-ORD ++ FLEX ++	*2.) Lateral cavaletti, on an elevation (Advanced)* ▸▸ *Trains the muscles of the shoulder girdle, chest muscles and fore limbs*
Forelimbs ③	CO-ORD ++ FLEX ++ STRENGTH +	*1.) Elephant trick with elevated hindquarters (Basic)* ▸▸ *Trains the muscles of the shoulder girdle, chest muscles and fore limbs*
	CO-ORD +++ FLEX ++ STRENGTH +	*2.) Lateral cavaletti with elevated hindquarters (Advanced)* ▸▸ *Trains the muscles of the shoulder girdle, chest muscles and fore limbs*
Forelimbs ④	STRENGTH +++ CO-ORD +++	*1.) Standing on a peanut ball/intermittent rocking (Basic)* ▸▸ *Full body workout*
Forelimbs ⑤	STRENGTH +++ CO-ORD ++ TACTILE ++	*1.) Pulling forward (Basic)* *Dog stands with his front feet on two half balls and is 'pulled' forward with food or hand target ('Hold')* ▸▸ *Dog lifts himself into shoulder suspension. Suitable as strength exercise for dogs with underdeveloped muscle in the chest and shoulder area*

Forelimbs 6	STRENGTH +++ CO-ORD ++	**1.) Looking down from an elevation (Basic)** Rear is elevated (ankle height on stable equipment) to allow the dog to look down ▸▸ This exercise is also a strength exercise for the entire trunk, chest and shoulder muscles
	CO-ORD +++ STRENGTH +++ TACTILE ++	**2.) Looking down double elevation (Advanced)** Standing on two half balls with the front feet increases the co-ordinative effect and contains a tactile component
Forelimbs 7	STRENGTH +++ CO-ORD +	**1.) Weight shift to the side (Basic)** Rear elevated (platform), front weight shift by means of hand touch or luring (no movement of the limbs) Isometric exercise with a lot of force intensity ▸▸ Training of trunk, chest and shoulder muscles
	STRENGTH +++ CO-ORD +++ TACTILE ++	**2.) Weight shift to the side/front elevated (Advanced)** Standing on two half balls with front legs to increase co-ordination and to set a tactile stimulus
	STRENGTH +++ CO-ORD ++ FLEX I I	**3.) Weight shift to the side/wobbly (Professional)** Shifting weight on a wobbly surface by looking in the direction of the side to be exercised Additionally trains body perception
Forelimbs 8	FLEX +++ STRENGTH ++ CO-ORD ++ TACTILE +	**1.) From paw to paw (Basic)** The dog is placed on a platform or board at the back and uses his front feet to step from one balance pad sideways to another ▸▸ Training of outer and inner shoulder blade muscles, plus chest muscles Also suitable for dogs who are very tight in the front

Forelimbs (9)	STRENGTH +++ CO-ORD ++	*1.) Push-ups (Basic)* *Dog stands with slightly elevated rear and performs an up and down movement with his head. Focus on the upper arm muscles (M.biceps brachii)*
	STRENGTH +++ CO-ORD +++ TACTILE +	*2.) Push-ups (Advanced)* *A more difficult variation is to place the back feet on a wobbly surface (e.g. balance pad). Trains additional co-ordination and is more strenuous than working on solid ground*
Forelimbs (10)	FLEX +++ CO-ORD ++	*1.) Front leg forward (Basic)* *Dog is stationed in a standing position with back feet on a low, firm platform. The front legs are stretched forward* ▸▸ *Active stretching of the M.Latissimus dorsi and the Musculus serratus (chest part) and the Musculus trapezius (chest part). Increased stretching of the outer shoulder area occurs when the front leg is moved to the side over the middle of the body.*
	FLEX +++ CO-ORD +++ TACTILE ++	*2.) Front leg forward from a rear elevation (Advanced)* *In order to challenge co-ordination, you can progress the exercise by placing the back feet on two half balls*

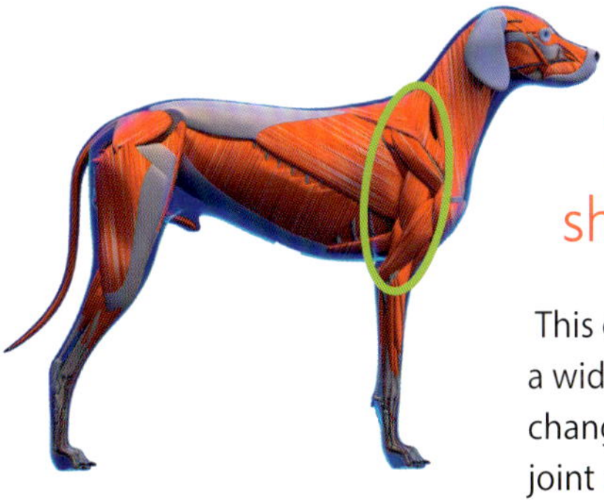

Exercise Band 1

+++ STRENGTH

+ TACTILE

1. Isometry with hand on shoulder (Basic)

This exercise is ideal for dogs of all ages and with a wide variety of problems, as it does not require a change in the length of the muscle and, therefore, no joint movement is required. The only important thing is to make sure that your dog is standing correctly and looking forward so the exercise has maximum effect.

1. This is where it helps if your dog is not necessarily in training mode, as it may be easier for him to stand still. However, you can also put your dog on a raised platform to encourage him to stay in position. At first, reward him for standing.
2. Start to move your hand. Click the dog for standing still, making sure he is not looking at your hand. If he is distracted, move your hand a very small distance. Alternatively, you can help him by using your verbal cue, "hold" (see Basics of Target Training, page 30). However, you will need to fade this, otherwise you will be missing a hand for handling.
3. If the dog tolerates the movement of your hand, gradually approach his shoulder blade. Click and reward each step.

The dog should stand parallel, both in front and behind.

4. Now place your hand on your dog's shoulder blade. This is the time when the consistent preparatory work pays off, as your dog should be happy to accept this. If he fidgets, go back to step 3.
5. Build up pressure slowly. Mark every counter-pressure with a click and reward generously.
6. Train to the point where the dog is happy to tolerate gentle pressure. When you release the pressure, you should feel as though the dog is 'falling' in your direction.
7. Now train on the other side.

STRENGTH +++

CO-ORD ++

Exercise Band 1

2. Isometry with a hand on shoulder/wobbly (Advanced)

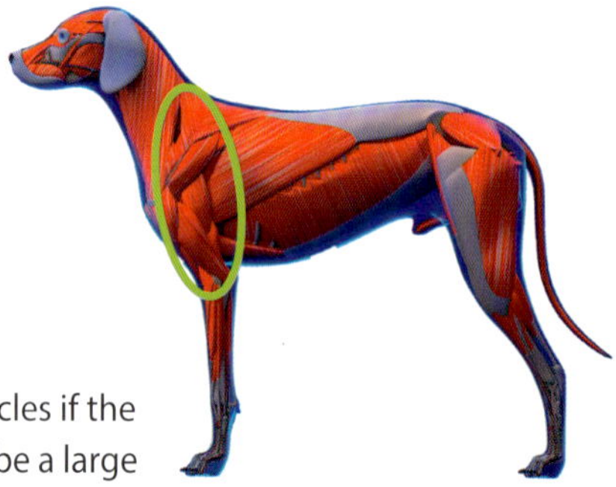

You can increase the effect on the shoulder muscles if the dog is standing on a wobbly surface. This could be a large air cushion, an old mattress, a trampoline or something similar. The dog should stand with his whole body on the wobbly surface.

The effect will be more intensive if only the front limbs or only the leg to be trained, is positioned on such a surface. Remember to place the rest of the dog on a firm, flat elevation, otherwise it will not be so effective.

Again, make sure the dog is standing parallel.

Equipment: TOGU Dynair Extreme.

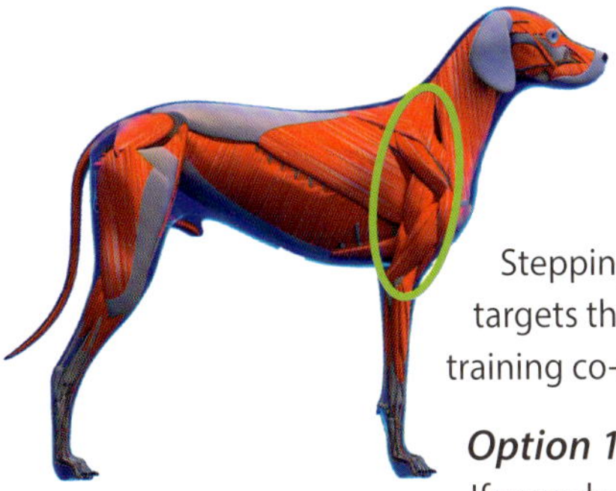

Exercise Band 2

1. Lateral cavaletti (Basic)

++ FLEX

\+ CO-ORD

Stepping over a cavaletti pole, from the side, not only targets the shoulder muscles, it is also a great exercise for training co-ordination.

Option 1 – Walking sideways

If your dog is already able to walk sideways, or knows the *elephant trick* (see page 75), practise two or three times, without the cavaletti bar. Now position the cavaletti on the floor, and ask him to repeat the behaviour. Reward generously if he steps over the pole. You can then progress to raising the pole gradually. The end position should be no higher than your dog's hock.

Option 2 – paw target

If your dog knows the paw target (see page 34), you can work with two paw targets, which you place to the right and left of the pole. Start with one target and place it next to the pole. Position the dog on the other side so that if he steps over, his front foot will hit the target. Click as he lifts his leg over the pole, and then reward. This exercise targets the forehand; if you want to train the whole body, the dog needs to step over with both his front and back feet.

Option 3 – Hand target

1. If the dog does not know the paw target, or sideways walking, you can make use of the hand target (see page 30). To start, help the dog over the cavaletti bar by offering a hand target next to the pole. If he steps over the pole with one foot, click and reward. Then walk a little further from the pole and reward the next step. At this point, a forward step rather than a sideways step is acceptable, so the dog gets the idea of stepping over the pole.

2. Now you want the dog to step over the pole sideways. You can achieve this by changing the position of your hand. As you start the exercise, position yourself so you walk backwards and guide your dog as he walks forwards, to step over the pole.

It helps if you stand in front of the dog for this exercise.

Equipment: Dog-Stepper.

FLEX ++

CO-ORD ++

Exercise Band 2

2. Lateral cavaletti on an elevation (Advanced)

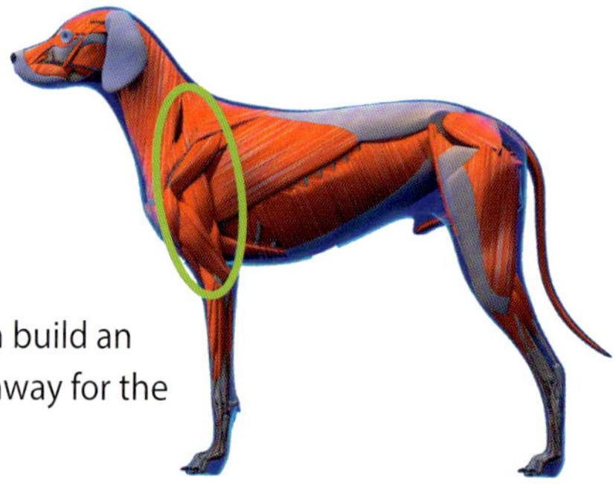

To increase the co-ordination challenge, you can build an elevation section under the pole, forming a pathway for the dog's front legs.

1. Your dog should first know how to step sideways with his front legs on an elevation. Practise this behaviour before progressing to the cavaletti bar.
2. Position the pole next to the dog so he steps over it on to the elevation. Click and reward generously.
3. Gradually position the pole towards the middle of the elevation. Now the dog has to step sideways across the floor, step on to the elevation, over the pole, back on to the elevation, and then back to the floor.
4. If this goes well, and the dog is enjoying the challenge, you can use several poles and vary the spacing. The bars should never be higher than your dog's hock. Observe closely to ensure the dog is moving smoothly, and always check that the poles are safe and cannot roll away if the dog steps on them.

The front elevation is used as an orientation aid for the dog.

Equipment: Dog-Stepper and TOGU Balance Bar.

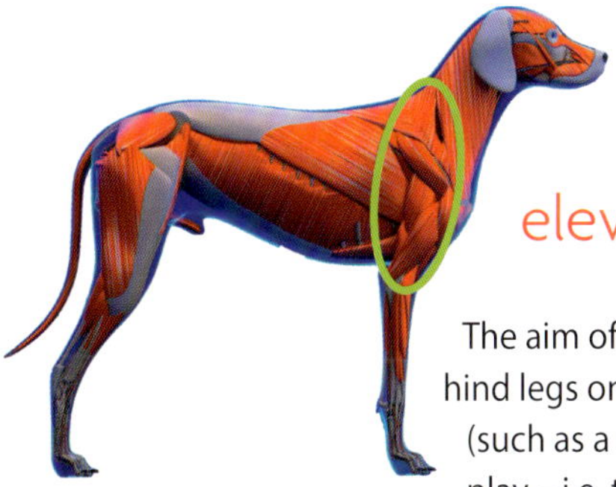

Exercise Band 3

1. Elephant trick with elevated hindquarters (Basic)

++ FLEX

++ CO-ORD

+ STRENGTH

The aim of this trick is to keep your dog stationary with his hind legs on an elevation, which is wobbly or filled with air (such as a balance cushion). Then movement comes into play – i.e. the dog walks a circle with his front legs in one direction and then in the other. The back legs move only minimally and remain on the elevation.

Option 1 – Shaping

You can shape the dog so that he places his back feet on an elevated surface.

Option 2 – Forwards

1. You need to work strategically so the dog makes the fewest mistakes, which means you can be quick to reward. First of all, find a confined area, e.g. a hallway, or build yourself an alley with the help of overturned chairs. Make sure there is sufficient width for the dog to traverse the alley comfortably. Arm yourself with a clicker and treats, and place your chosen elevation, e.g. a balance cushion, at the start of the alley. Now ask the dog to join you.

2. If the dog shows the behaviour in a variety of places, you can introduce a verbal cue for the trick.

3. Now, when the dog is in position, with his back feet on the balance cushion, place yourself directly in front of him and offer a hand target at the level of his mouth. Repeat this at least five times.

4. Start to move slowly in a circle around the balance cushion. At first, only minimally, so the dog only has to stretch his head to the side. Click and reward this.

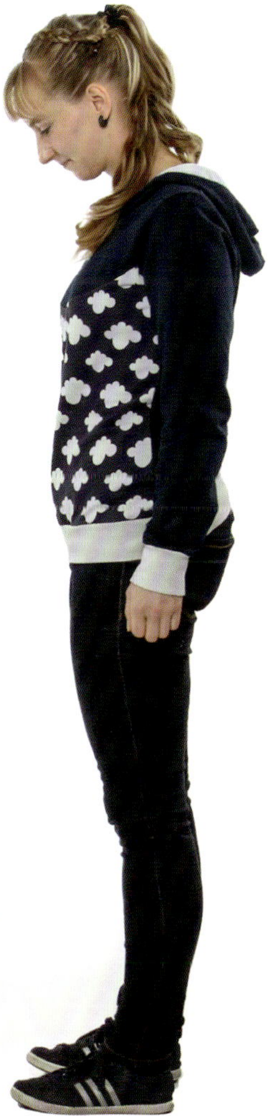

Here you can see the starting position.

Equipment: TOGU Dynair ball cushion.

5. Now, move a little further away from the dog, but in the same direction as before. Hold the hand target centrally in front of your body. Your dog will now need to turn slightly. It is important that you click the movement of the forehand, as the hindquarters should remain on the elevation. Repeat to build on success.
6. If this exercise works in one direction, then go to the other side and proceed in the same way as outlined in steps 7 and 8.
7. If your dog doesn't make any effort to follow you in both directions, only reward him after two or more steps. Vary the direction and reward rate.
8. Now you can introduce a verbal cue for each direction, or you can continue working with a hand target.

Option 3– Backwards

1. If the dog knows how to walk backwards, work with this, creating an alleyway with an elevation at its end.
2. Send the dog backwards on to the elevation, and reward generously. Repeat this until he starts to search for the elevated surface with his back feet
3. Continue with step 5 of option 2

FLEX ++

CO-ORD +++

STRENGTH ++

Exercise Band 3

2. Lateral cavaletti with elevated hindquarters (Advanced)

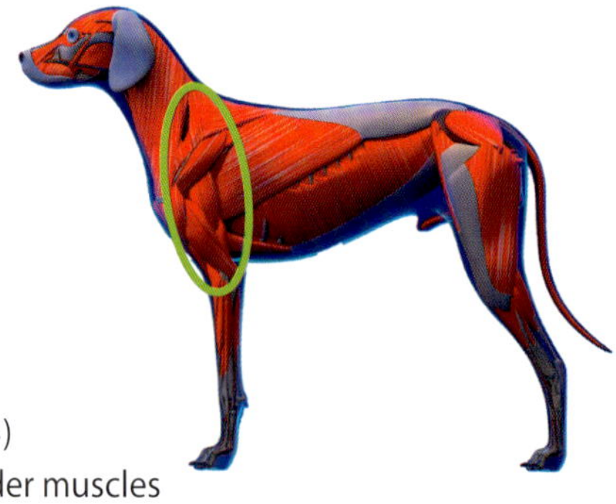

This extension of the *elephant trick* (see page 75) increases the training effect on the dog's shoulder muscles as, when he steps over the cavaletti pole, he uses his forehand more consciously and he also puts his feet wider apart.

1. Place a cavaletti pole directly on the elevation and secure it against rolling away.
2. Then fetch your dog together with a clicker and treats.
3. Start on the side without the cavaletti pole and ask for the elephant trick .
4. Move with the dog sideways around the elevation until you reach the cavaletti pole. The dog may need a little help here, so offer a more prominent hand target. If he hesitates, wait and give him time to think. Attempts in the right direction should be generously rewarded.

5. If the dog steps over the pole, repeat this several times by finishing the circle.
6. If this works without any problems, you can raise the level of the pole.
7. Now work on the other direction.
8. Once the dog is fluent in both directions you can alternate them, and work with several poles placed at irregular intervals around the circle. Voila! You have now put together a very intensive shoulder workout for your dog. Do not overdo it with too many repetitions

Here you can see how the chin target can be used to bring more calm into the exercise for over-motivated dogs.

Equipment: TOGU Brasil on TOGU Dynair ball pillow, in between cavaletti pole.

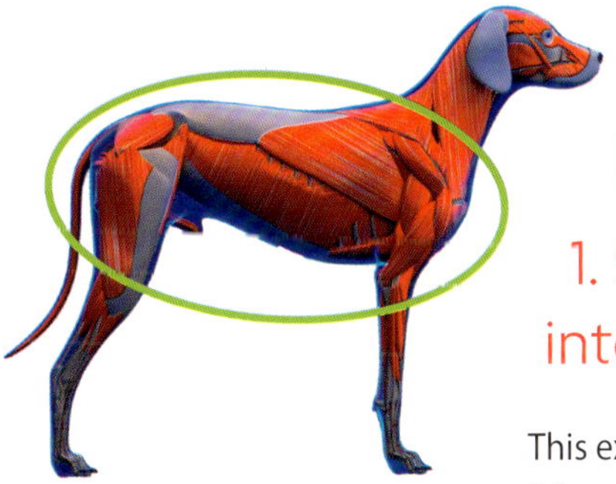

+++ CO-ORD

+++ STRENGTH

Exercise Band 4

1. Standing on a peanut / intermittent rocking (Basic)

This exercise is not difficult to train in itself, but it is extremely strenuous for the dog. Therefore you should only work on it if your dog appears completely comfortable and shows no sign of muscle trembling.

1. First of all, make sure the dog does not have unsupervised access to an unsecured peanut ball at any time, as this could cause an injury.
2. Initially, many dogs are reluctant to get on the wobbly peanut ball. Instead of letting the dog bounce directly on to the ball, it is a good idea – especially for hesitant dogs – to use a stool. Start by offering a hand target and encourage the dog to sit on a stool. Praise him for doing so. From there, it will be easier for him

to get on to the peanut ball. To begin with, the dog may feel more comfortable if he is sitting or lying down, in which case, work gradually towards the standing position, giving the dog time to get used to the equipment.

3. Now withhold your reward until the dog is standing still. If he is restless, or thinks he should offer different behaviours, you can use the 'hold' to show him that it's all about relaxed standing.

4. Wait until the dog has perfected a relaxed stand, and then start to move the peanut ball gently back and forth. Observe the dog closely and adjust the movement so that he is always able to maintain his balance. He should be able to stand still on the peanut ball, despite the intermittent rocking motion.

5. Test at 15-second intervals to ensure the dog is still comfortable with the exercise.

The peanut ball should be used on a non-slip surface.

Equipment: FitPaws studded peanut ball.

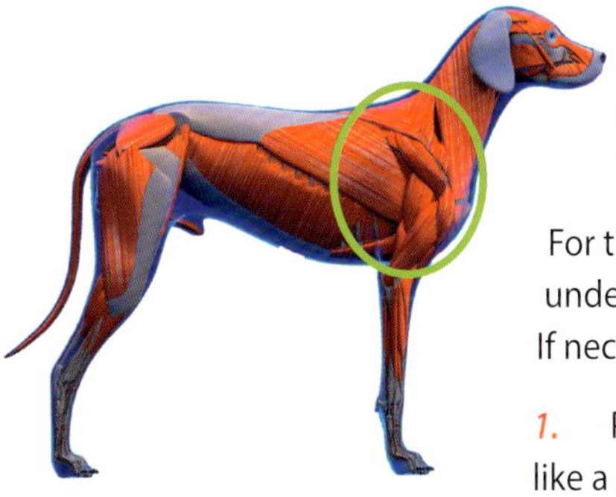

Exercise Band 5

1. Pulling forward (Basic)

++ CO-ORD

+++ STRENGTH

++ TACTILE

For this exercise, it helps if the dog has a full understanding of paw targets (see page 34). If necessary, refresh it beforehand.

1. Place two half balls in such a way that they look like a large front paw target. If necessary, you can click the dog for initial interest, then for putting one foot on a ball and, finally, for standing with one foot on each half ball.
2. 2. For this exercise it is essential that the cervical spine is horizontal and the dog remains in the desired position. You should, therefore, decide whether to work with a hand or chin target or, if necessary, with treats to show your dog the position first.
3. Offer the target quite close to the tip of the dog's nose and click the first touch. It is important that he stops on the half balls.
4. Now 'pull' the dog forwards by offering the target, in such a way that his neck and spine remain parallel to the ground. This will automatically activate the shoulder area.
5. The more you 'pull', the more weight the dog has to take up with his front legs, and the more effective the training is for the shoulder girdle. If you overdo it, the dog may step off the half balls. If this happens, just make the challenge easier.

The smaller the half balls, the greater the challenge, in terms of the co-ordination required.

6. Reward in the starting position; this creates an activating effect as you switch between 'relaxed standing' and 'pulling'.

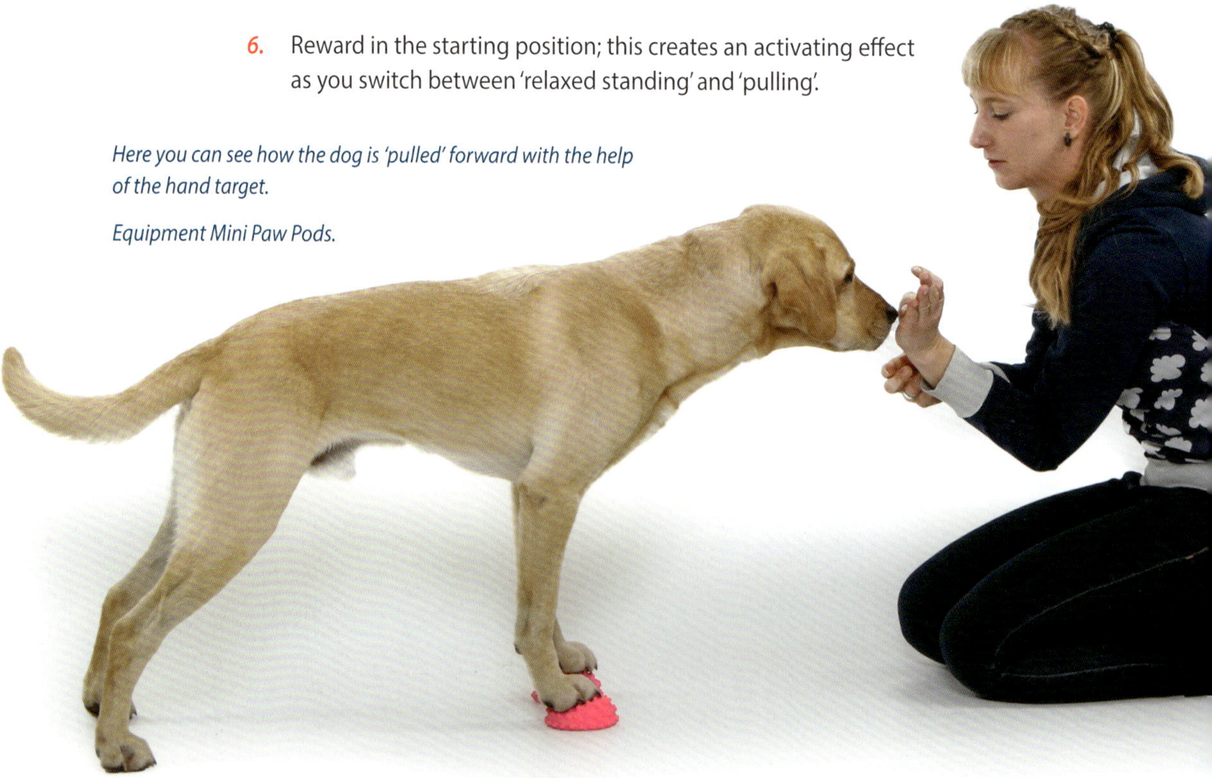

Here you can see how the dog is 'pulled' forward with the help of the hand target.

Equipment Mini Paw Pods.

CO-ORD ++

STRENGTH +++

Exercise Band 6

1. Looking down from an elevation (Basic)

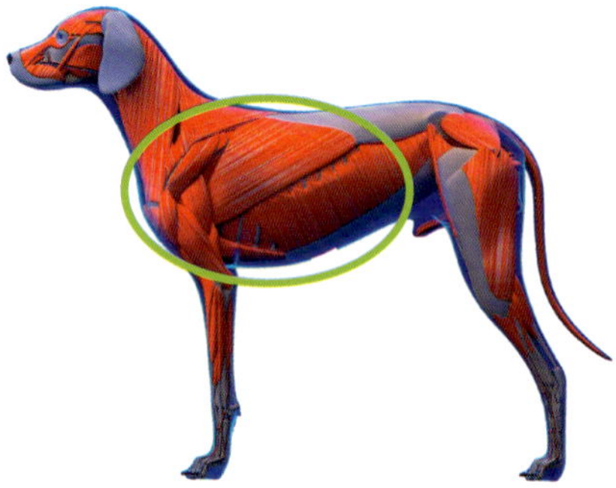

For this exercise, it helps if the dog has a full understanding of paw targets (see page 87). If necessary, refresh it beforehand.

1. You will need a platform that is no higher than your dog's hock, a clicker, treats and Post-it notes.
2. Refresh the nose target with a Post-it.
3. Guide the dog over the stool from behind. Reward him several times for standing calmly with his back feet on it.
4. Offer the dog a Post-it as a target, which he should touch with his nose. Click and reward. Dogs who like using their paws, will, generally, not attempt to do this when their hindquarters are raised. Nevertheless, you should offer the target at your dog's eye level at first.

5. Once you have established the nose touch, you can gradually lower the target towards the floor. Click and reward the intermediate steps.

Tip: If this training is new to your dog, lead him over objects that are very low to the ground, and then, gradually, increase the height.

6. Now place the target at a slight distance from the dog's front feet so he can reach it comfortably with his nose, without having to move from a standing position.
7. Be consistent and only mark nose touches on the target. Now the placement of the reward becomes significant. To encourage the dog to repeat the behaviour, don't feed him at target level. Hold the treat high enough to keep his head straight.
8. After he has eaten, the dog will look down towards the target. Click the touch and, again, reward with the treat higher up.
9. Add a verbal cue just before the dog restarts the exercise.
10. Gradually decrease the size of the target, but don't be in too much of hurry or you will run into trouble. If this happens, feed the dog directly on the target to rekindle his interest.
11. Now try to eliminate the target completely. Click the dog when he is at the lowest point of his downwards movement. He will automatically look up to get his treat, thus activating the shoulder girdle. If you would prefer the dog to stretch, he should hold the downwards position for at least 10 seconds.

Make sure the dog stands parallel and has a straight back line. The height of the hand target influences this significantly.

Equipment: Ikea stool (small).

CO-ORD +++

STRENGTH +++

TACTILE ++

Exercise Band 6

2. Looking down from a double elevation (Advanced)

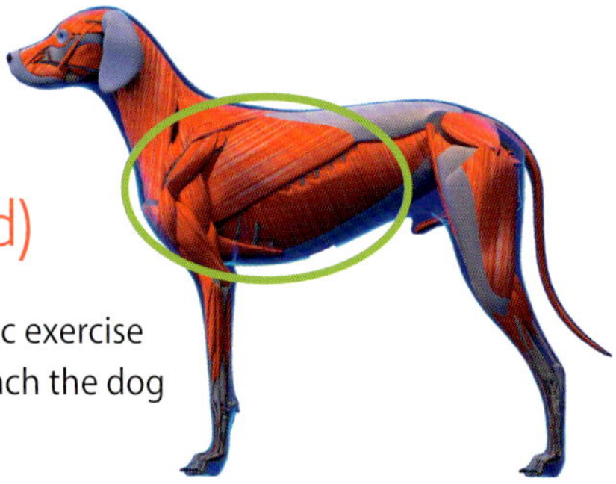

The co-ordination challenge involved in the basic exercise (see page 80) is increased if you, additionally, teach the dog to balance his front feet on two half balls.

1. Get ready with two half balls, plus the stool, which the dog already knows.
2. Proceed, before, by asking the dog to position his back feet on the stool. Reward several times by clicking and throwing the treat further away, so the dog has to re-position himself on the stool.
3. Now place two half balls in position for the front feet. Generally the dog will accept the new equipment relatively quickly, which means that you can move on to clicking him for standing on them.
4. If the dog is stepping on to the half balls, reward this several times by throwing the treat behind the stool so the dog has to leave the equipment to eat them. You can then mark and reward the dog when he re-positions himself.
5. If the dog is reliable in positioning himself on both sets of equipment, you can add the basic looking down exercise (see page 80).

Tip: If your dog is finding this exercise difficult, work on the basic exercise to relax the back: front and rear targets, (page 134). This involves positioning the dog on targets at the front and back and gradually increasing the distance between them. Initially, the front and rear targets should be trained separately.

The half balls for the front feet provide an additional challenge.

Equipment: TOGU half ball and Ikea stool (small).

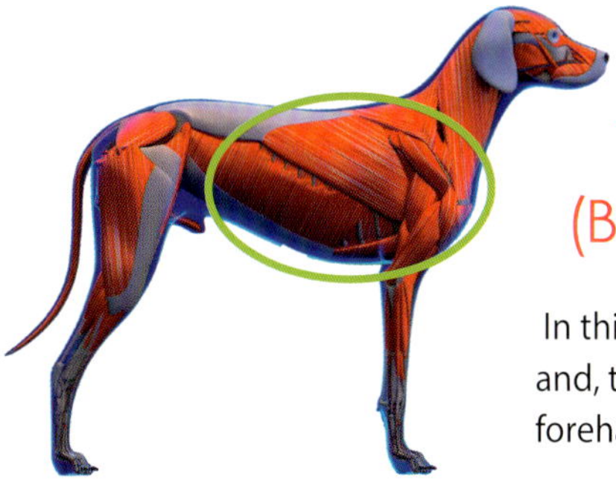

Exercise Band 7

1. Weight shift to the side (Basic)

\+ CO-ORD

+++ STRENGTH

In this exercise, the dog learns to lean to one side and, thus, to put more weight on the respective forehand.

1. Provide a small elevation. Send the dog to the target on the elevation by walking backwards (see *backing down an alley*, page 172).
2. Then click and reward for stepping on to the elevation, first with one back foot and then with both back feet.
3. Now use a hand touch or lure hand to encourage the dog to stretch to the side and, thus, take more weight on one front foot. You will need to make sure his spine is straight.
4. Let him hold this position for 10 to 15 seconds. If necessary, you can support him, using the chin target.

Tip: Some dogs will learn more easily if you build an alley, which provides a visual boundary. Lead your dog through the alley by walking backwards until you are past the elevation – this way he is more likely to step on to it. Alternatively, you can put your dog in the 'between' position (between your legs), and thus lead him to the stool.

Here the handler is standing, but it may help if you go down to the dog's level if he has a tendency to look up.

Equipment: Ikea stool (small).

CO-ORD +++

STRENGTH +++

TACTILE ++

Exercise Band 7

2. Weight shift to the side/ front elevated (Advanced)

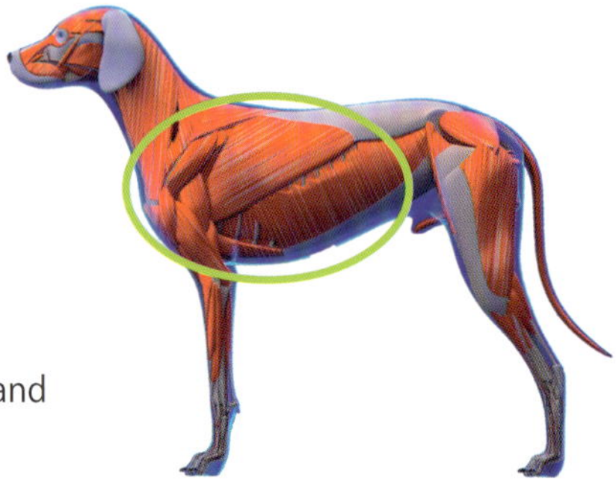

To increase the co-ordination challenge, ask the dog to step on two half balls with his forehand, and weight shift sideways.

Option 1 via giving paw

1. If your dog is able to give paws, first ask for his front paws, in turn, while he is standing with his back feet on the elevation.
2. Now the half balls come into play. First place one in front of the dog and ask for his paw. Allow him to put the paw in your hand once or twice, then place your hand on the half ball and gradually ease it away.
3. Repeat with the other paw.
4. Now throw the reward behind your set-up and wait to see whether the dog can re-set his position again. If necessary, you can help with your 'give paw' cue.

Here you can clearly see the weight shift from left to right.

Equipment: Ikea stool (tall) and TOGU half balls.

Option 2: via front paw target

1. If your dog knows the paw target, first refresh it without the equipment.
2. Then ask the dog to place his back feet on the stool. Reward him, throwing the reward away to the back so he has to start the exercise anew.
3. Now the paw target is added. Make use of the moment when the dog eats his treat and place the paw target at the appropriate distance in front of the hind paw target.
4. Click and reward the dog for placing one or both feet on the target.
5. From now on, he is only rewarded when both feet are in position.
6. Throw the treat behind him and replace the familiar paw target with the two half balls. Reward any paw action until you manage to get both feet on the half balls.

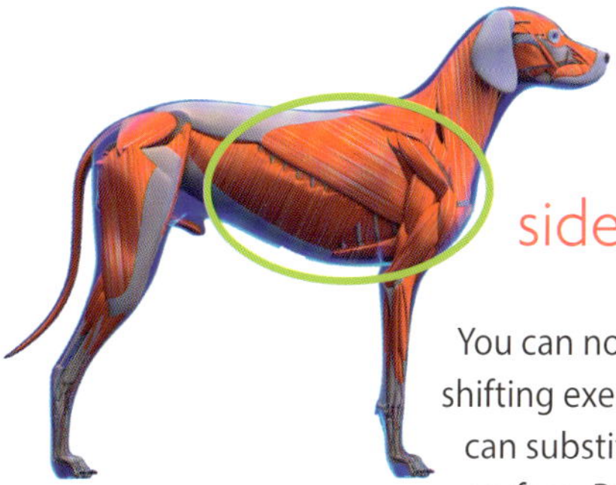

Exercise Band 7

3. Weight shift to the side/wobbly (Professional)

++ FLEX

++ CO-ORD

++ STRENGTH

You can now transfer the whole sideways weight shifting exercise to a wobbly surface. Alternatively you can substitute the stool for the back feet with a wobbly surface. Remember, the weight is on the forehand.

Here you can see the extent to which the handler's posture influences the weight shift of the dog.

Equipment: TOGU Dynair Extreme.

FLEX +++

CO-ORD ++

STRENGTH ++

TACTILE +

Exercise Band 8

1. From paw to paw (Basic)

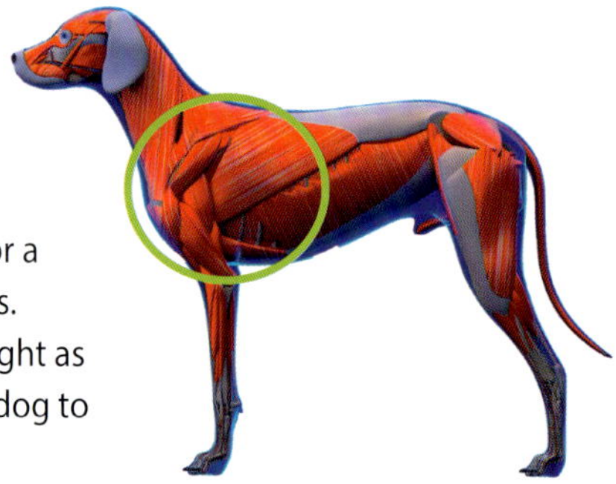

1. You will need a platform (e.g. dogstepper) or a board, as well as two balance pads/cushions. The platform should be about the same height as the balance pads and wide enough for the dog to stand on.
2. Refresh the paw target with a balance pad, and then put it away.
3. Now get the platform ready. This time, it is not a question of the dog standing on an elevation, but of his correct positioning. Guide the dog on to the platform from behind, click and reward him for standing on it. After the click, throw the treat behind the platform so he has to re-set his position.
4. Once the dog has found his position, place a balance pad in front of the platform, slightly offset to the side. By refreshing the paw target, the dog will probably show interest in the pad of his own accord, and stand on it with his front feet. If not, you can use your verbal cue to ask for the paw target, or your hand target as a guide. Mark this intermediate step, i.e. standing with his back feet on the platform and both front feet on the laterally offset balance pad – only once – and feed in this position.

Tip: You can also work with a rear target for the back feet, but a larger area has proved to be more suitable for inexperienced dogs.

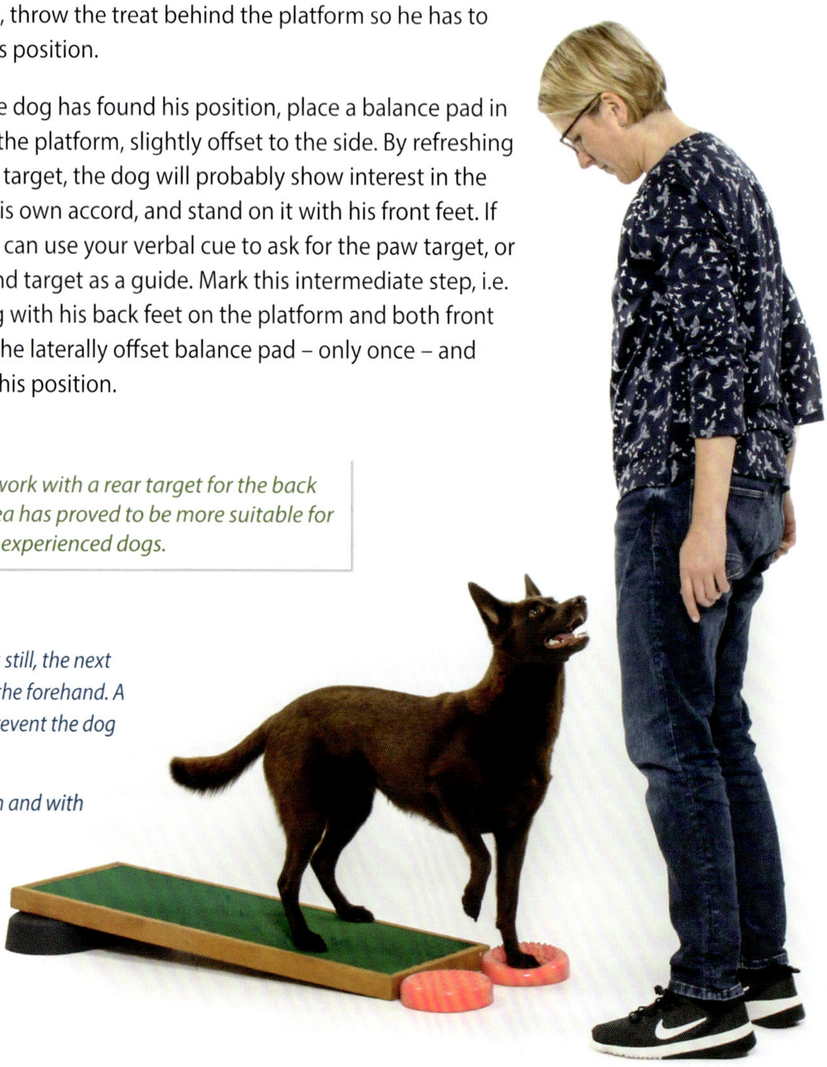

If the dog is able to keep his hindquarters still, the next step is to initiate sideways movement of the forehand. A sitting handler would be better here to prevent the dog looking up.

Equipment: TOGU Senso balance cushion and with underlaid board.

5. Now the second balance pad comes into play. Keep your clicker and treats at the ready when you place the second balance pad next to the first. Hopefully, the dog will try to step on to the second pad, and this is exactly what you want to mark with the clicker. Follow up by rewarding in position. If the dog is not interested in the second balance pad, encourage him verbally and give additional help him with a verbal cue or a hand target.

6. For maximum effect, make sure the dog remains in position on the board and does not move his back feet.

7. Switching back and forth between the pads is usually quite easy as most dogs seem to enjoy it. Initially, you can help with a hand target or cue the paw target.

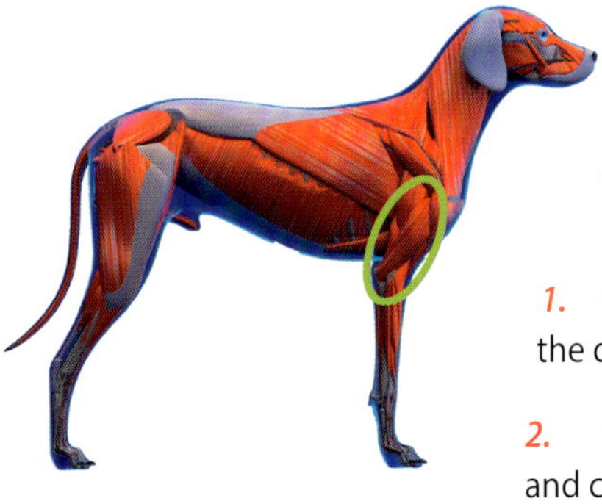

Exercise Band 9

1. Push-ups (Basic)

++ CO-ORD

+++ STRENGTH

1. For this exercise, you need a small elevation for the dog's back feet.

2. Lead your dog over the elevation from behind and click and reward for placing his back feet on it.

3. Once the dog has his back feet in position, you can start working on his head movement. You can do this with a hand target, a nose target using a Post-it, or you can lure with food. It is important that the dog works calmly and that his elbows show a slight bend.

Option 1 – Hand target

1. Ask the dog to touch your hand at a height that means he has to bend his elbows slightly.

2. Click for the touch, but reward him further up, so that he no longer bends his elbows and keeps his head in a normal position. From there, the dog can start the exercise again.

3. Offer him the hand target again and train the whole behaviour 10 to 15 times.

Option 2 – Post-it

Proceed as for option 1, except you will be using a Post-it in your hand, or working with the chin target/hold.

Option 3 – Food

You can lure the movement with food, making sure you work accurately and proceed in a calm manner.

Tip: If your dog has difficulty hitting the elevation, start by working with several dogstepper elements. When he has learnt to find the elevation, you can gradually reduce the number of elements. You can also work with a visual boundary, for example, working along a wall or setting up cavaletti poles on either side.

Using a hand target, or lure hand, guide the dog's head upwards, and downwards, so there is a slight, repeated bend at the elbow.

Equipment: Dogstepper.

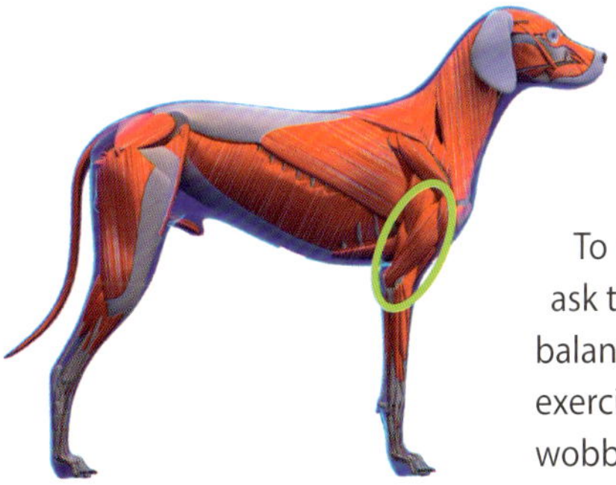

Exercise Band 9

2. Push-ups/wobbly (Advanced)

+++ CO-ORD

+++ STRENGTH

\+ TACTILE

To make the exercise more difficult and to intensify the effect, ask the dog to stand with his back feet on a wobbly surface, e.g. · balance cushion. To do this, proceed in the same way as with the b exercise (page 87), first rewarding the dog for standing calmly on the wobbly surface before working through the rest of the exercise.

Here you can see the more difficult push-up version. Equipment: TOGU Brazil.

FLEX +++

CO-ORD ++

Exercise Band 10

1. Front leg forward (Basic)

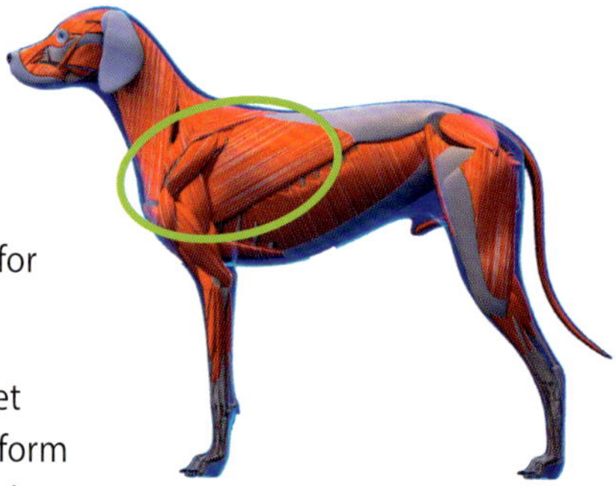

1. For this exercise, you need a small platform for your dog's back feet.
2. Lead the dog from behind with a hand target and reward him when he stands on the platform with his back feet. Throw the treats away to the back so that he learns to re-set his position on the platform.
3. Now sit down in front of the dog.
4. Hide a treat in your fist and offer it to him at paw height.
5. Click for any paw action – do not click a nose touch.
6. Strengthen the paw action by marking and rewarding the behaviour you want.
7. Now turn your fist so that the back of your hand is facing down. Reward a touch of your hand.
8. Remove the treat from your hand and click and reward for touching the empty fist. Make sure you feed from the other hand.
9. Shortly before the dog starts the next 'paw stroke', open your fist so that he touches your flat palm.

In this exercise the elbow should be supported, and no counter-support should be given if the dog pulls back.

Equipment: Dogstepper.

10. Click and reward this several times and introduce a verbal cue for this paw. For example, "tip" is a good choice, as it contains an 'i', as does 'right'. You can then call the left paw "tap".

11. Now train the other paw.

12. You can ask for both paws alternately to get an active stretch. Also vary the length of time the paw is in your hand from very short to long duration. If the dog gives you his right paw, more weight will lie on his left forehand and vice versa. You can, therefore, focus on the side that needs more work.

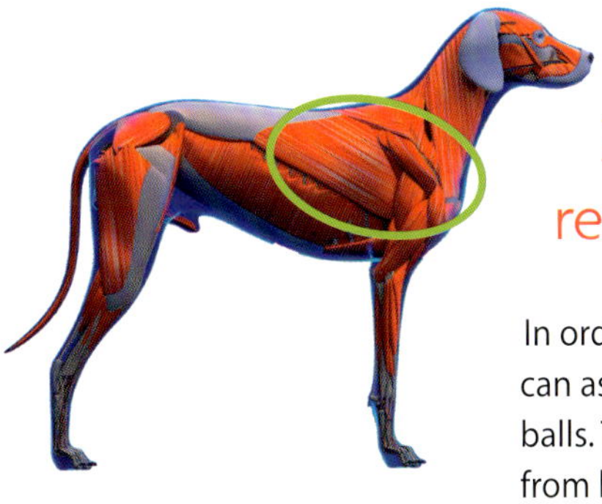

Exercise Band 10

2. Front leg forward from a rear elevation (Advanced)

+++ FLEX

+++ CO-ORD

++ TACTILE

In order to increase the co-ordination challenge, you can ask the dog to place his back feet on two half balls. To do this, lead him over the half balls, again from behind, and reward him first for placing his back feet on them. If he feels comfortable here, you can then ask the forward movement of his front paws.

In this advanced version, the dog also needs to co-ordinate his hindquarters.

Equipment: TOGU half balls.

IV. EXERCISES FOR THE FRONT THIGH MUSCLES

Front thigh muscles		
Category:		***Hind limbs***
Training goal:		*Strengthening and co-ordination*
Recommended:		*For dogs with under-developed thigh musculature, for stabilisation during sporting activities and in the case of existing conditions (e.g. patella luxation)*
Muscular function:		• *The quadriceps femoris muscle with its four parts takes over the stretching of the knee, the fixation of the limb and is involved in the flexion of the hip joint* • *Muscles that are trained (depending on the exercise) are the M.sartorius, M.tensor fasciae latae and the M.iliopsoas – all these muscles play a major role in swinging forward the fore limb.*
Front thigh muscles ①	STRENGTH ++ CO-ORD + FLEX +	***1.) Sit-Stand-Sit (Basic)*** ▸▸ *The quadriceps group, as well as the ilium-lumbar muscle, is trained*
	STRENGTH +++ CO-ORD ++ FLEX +	***2.) Sit- Stand-Sit/front elevated (Advanced)*** ▸▸ *Training of the quadriceps group and the ilium-lumbar muscle with greater exertion than in the basic exercise*
	STRENGTH +++ CO-ORD +++ FLEX ++	***3.) Sit-Stand-Sit on two peanut balls of different heights (Professional)*** ▸▸ *Training of the quadriceps group with simultaneous balancing*

Front thigh muscles ①	STRENGTH +++ CO-ORD +++ FLEX ++	**4.) Knee bend on a peanut ball (Professional)** ▸▸ Thrust steeply upwards; very strenuous but effective workout for the front thigh muscles
	STRENGTH ++ CO-ORD ++ FLEX +	**5.) Sit on a peanut ball (Professional)** ▸▸ Here, the back, the abdominal and the ilium-lumbar muscles are also used
	STRENGTH +++ CO-ORD ++	**6.) Kick-back on a peanut ball (Professional)** ▸▸ Training of the entire hind limbs, the back and the ilium-lumbar muscles
Front thigh muscles ②	STRENGTH ++ CO-ORD +	**1.) Lifting the buttocks (Basic)** ▸▸ Small movement, but very strenuous for the front thigh muscles
	STRENGTH ++ CO-ORD +	**2.) Crawling (Advanced))** ▸▸ Corresponds to footballers' warm-up – running with bent knees. Very strenuous
Front thigh muscles ③	FLEX +++ CO-ORD ++ STRENGTH +	**1.) Narrow cavaletti (Basic)** ▸▸ Lots of joint movement (knee and hip) and good strength training for the thigh area
Front thigh muscles ④	STRENGTH +++ CO-ORD ++ FLEX ++ TACTILE +	**1.) Climbing stairs (Basic)** ▸▸ The slower the exercise is done, the more effective it is. Great strengthening of the front thigh muscles plus improvement of the thrust from the hindquarters

Front thigh muscles ④	STRENGTH +++ CO-ORD ++ FLEX ++ TACTILE +	**2.) Stepping up with hind legs/wobbly (Advanced)** ⏩ *Perfect training if one side needs more work; similar effect to climbing stairs*
Front thigh muscles ⑤	STRENGTH +++ CO-ORD + TACTILE +	**1.) Lunge forward (Basic)** ⏩ *Power thrust but no joint movement*
Front thigh muscles ⑥	STRENGTH +++ CO-ORD +++	**1.) Weight shift on a wobble board (Basic)** ⏩ *Isometric exercise, trains the entire thigh and back muscle.*
Front thigh muscles ⑦	STRENGTH +++ CO-ORD ++	**1.).Bounce to a hand touch (Basic))** ⏩ *Power thrust. Avoid for dogs with spinal problems*
	STRENGTH +++ CO-ORD +++ TACTILE +	**2.) Bounce to a hand touch/wobbly (Advanced)** ⏩ *Increased effect through balancing*
Front thigh muscles ⑧	STRENGTH +++ CO-ORD ++	**1.) Mountain goat (Basic)** ⏩ *In addition to the front thigh muscles, the back thigh muscles, abdominal and back muscles are also trained*
	STRENGTH +++ CO-ORD +++	**2.) Mountain goat /wobbly (Advanced)** ⏩ *Balancing poses an additional challenge*

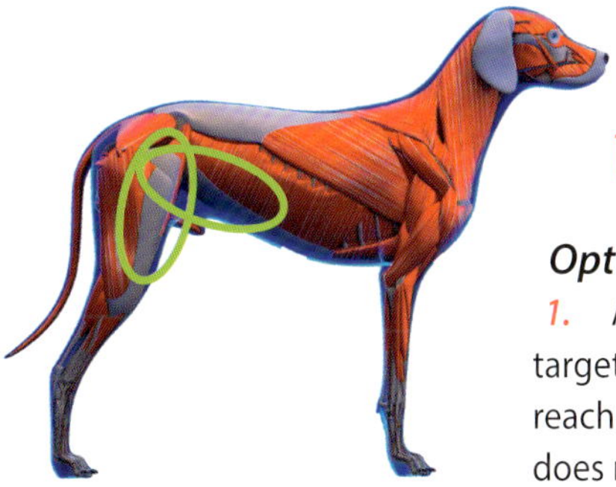

Exercise Band 1

1. Sit-Stand-Sit (Basic)

+ FLEX

+ CO-ORD

++ STRENGTH

Option 1:

1. Ask the dog sit in front of you. Offer a hand target at such a distance that he has to stand up to reach it. Make sure that he only has to stand up and does not have to take any further steps. Click and reward. Repeat this exercise several times.
2. If the dog goes from sit to stand, just as you wish, introduce a verbal cue before you offer your target hand.
3. Now alternate between sit and stand, making sure the dog is maintaining the correct posture and body tension.

TIP: If your dog steps forward and you don't manage to catch the 'pure' stand, practise on an elevation, such as a small, raised platform. In this way, the dog has to go direct from sit to stand, otherwise he would step off the elevation.

Option 2:

1. Ask the dog to stand in front of you, and then to walk backwards. Reward him for this.
2. Now ask the dog to lie down in front of you. Here, it is important to have good timing. Ask the dog to walk backwards and click at the exact moment that he lifts his bottom off the ground and straightens his hind legs. You are marking, and rewarding, the first movement of getting up – the start of walking backwards.
3. Repeat this several times.
4. If this is going well, introduce a verbal cue.
5. Now alternate between sit and stand, making sure the dog is maintaining the correct posture and body tension.

Here, the platform helps as a spatial limitation for the dog.

Equipment: Dogstepper.

FLEX +

CO-ORD ++

STRENGTH +++

Exercise Band 1

2. Sit-Stand-Sit/front elevated (Advanced)

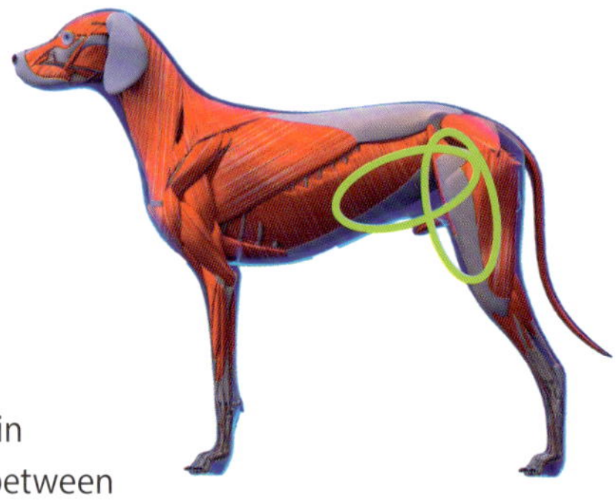

1. First of all, the dog needs to get in position with both front feet on an elevation. Depending on the dog, it may help to start in the heel position, or with the dog walking between your legs. As a visual aid, an alley can be helpful for more hesitant dogs. Initially, the elevation should be minimal.
2. Click for the dog showing interest in the elevation.
3. Then click only when he moves on to it, first with one front foot, then with both.
4. It is not difficult for the dog to stand correctly in this position; it is much harder to transition to a sit. Ask for the sit and reward generously if he responds correctly. If not, use a hand target to help him into position. Some dogs are unaccustomed to sitting at different heights. Be patient and click the first few attempts if the dog is getting the idea.
5. If the dog is able to change position with his front feet on a small elevation, you can gradually start raising the height. The higher the forehand, the more work the hindquarters have to do when changing position. You can also raise the forehand on to a wobbly surface to improve forehand co-ordination.

Tip: If your dog finds it difficult to switch from standing to sitting, the elevation may be too small. In this case, work with an elevation that is longer in length, which will allow him to move forward.

Pay close attention to the correct execution of the movement transitions.

Equipment TOGU Aero-Step One.

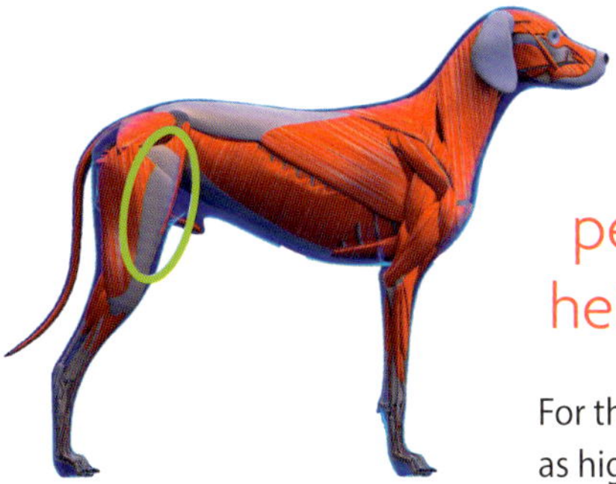

Exercise Band 1

3. Sit-Stand-Sit on two peanut balls of different heights (Professional)

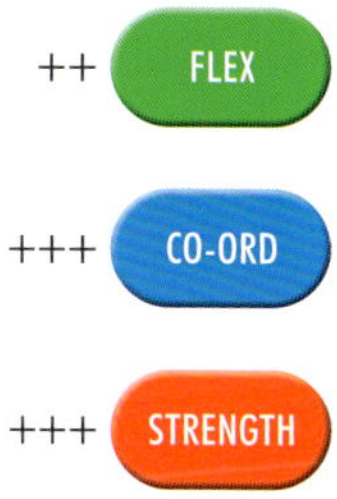

For this exercise, the front peanut should be at least as high as the rear peanut, otherwise you would be training the forehand rather than the hindquarters. Dogs find balancing on two peanuts easier than on one, and this is especially the case with large dogs.

1. Start with just one peanut and replace the front peanut with a stable elevation, at about the same height. Make sure that the peanut cannot move while the dog is working on it.
2. Lead your dog up to the peanut from behind. If he shows interest in it, click and reward him generously.

3. If he fails to engage with the peanut, help him with a hand target. Offer it so that the dog can stand with his paws on the ground and only his head is above the peanut. Then move forward, a step at a time, until he places first one front foot, then both, on the peanut. Reward each step generously.

Initially work with only one wobbly surface, positioned at the rear. When the dog is confident, bring the second wobbly surface into play.

Equipment: FitPaws FitBone and FitPaws Donut with holder.

4. This is the most important step: the dog has to bring his back feet up on to the peanut. Support him verbally and physically, if he needs it. Wait until he can stand in a relaxed manner before progressing.

5. Now place a stool in front of the peanut, and encourage the dog to step forward so his front feet are on the stool, and his back feet are on the peanut. Reward him for this, and keep practising until the dog is happy in this position.

6. Now you are ready to add the front peanut. The dog needs to start at ground level, so bring him back down and ask him to wait while you change the set-up. In principle, all you need to do is place the second peanut in front of the first and invite the dog to climb on and take up his position on both peanut balls.

7. Click and reward this in detail. The dog should feel really comfortable before you ask for more.

8. When he is ready, ask for the stand and sit alternately. When doing this, make sure his hindquarters remain on the rear peanut. If he slides forward to a sit, he is not ready for this exercise. If this is the case, you need to go back a step and practise a relaxed stand on the two peanuts, or even go back to using a stool for his front feet. If the dog is very experienced you could ask him to reverse back so he is in the correct potion, but be very careful not to let him slip.

9. Repeat the change from sitting to standing several times, but make sure that there are no signs of fatigue, such as muscle trembling or panting.

FLEX ++

CO-ORD +++

STRENGTH +++

Exercise Band 1

4. Knee bend on a peanut ball (Professional)

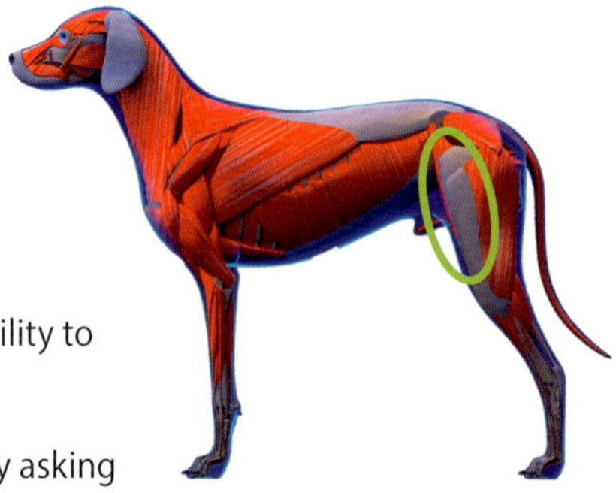

This exercise is about the dog developing the ability to thrust upwards from his hindquarters.

1. Get your dog in the mood for the exercise by asking him to do several sit-stand changes or, if he is familiar with this, the change between sit and 'beg'.

2. Now the peanut ball comes into play. It should be neither too big, nor too small. The ideal size is when the dog is standing with his front feet on it, and his spine is at a 45-degree angle to the horizontal. You can also position the peanut ball lengthways and hold it at an angle to get the right height.

3. Ask the dog to stand on the peanut ball with his front feet only. Stand opposite to him as this prevents excitable dogs from jumping on to the peanut. Cautious

dogs may need more help. You can use a high value treat for encouragement, or you can use liver paste, which is high value, but the requirement to lick also has a calming effect.

4. If the dog stands with his front paws on the peanut, reward this starting position several times.

5. Now ask for a sit. Make sure you mark and reward initial attempts. The dog is unlikely to go into a full sit straightaway.

6. Work on a gradual basis. Observe the dog closely to ensure he maintains the correct tension in his body throughout the exercise, and does not fidget with either his back or front feet.

7. Now ask the dog to stand from the sit by using a verbal cue. Alternatively, you can use a hand target, raising it upwards in an imaginary line.

8. Once your dog has understood the exercise, ask him to do it several times in succession.

Here the dog has to push up more steeply than in the previous exercise.

Equipment FitPaws pimpled peanut ball.

Tip: It may help to provide a small platform for the back feet, which acts as a visual limitation for this exercise.

FLEX +

CO-ORD ++

STRENGTH ++

Exercise Band 1

5. Sit on a peanut ball (Professional)

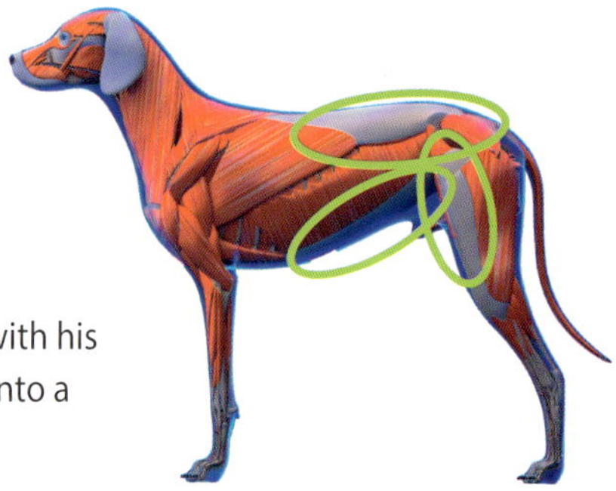

In this exercise, the dog sits on the peanut ball with his hindquarters bent and pushes himself forward into a standing position.

1. For this exercise, you need a peanut ball secured on a stand, or an assistant to keep the peanut in position.
2. Lead the dog over the peanut from behind, keeping back feet on the peanut and his front feet on the floor. The dog's spine should be in a horizontal position, or sloping down towards the hindquarters.
3. First reward the starting position – at this stage the dog is still standing.
4. Now ask him to sit, making sure his rear end remains on the peanut. Reward this position several times.
5. The dog will sit with his front legs quite close to his back legs. Now your task is to encourage him to 'pull' forwards. You can do this very effectively using a chin target.
6. Again, click and reward every little step. The end product of the exercise is a dog that keeps his rear end on the peanut but, otherwise, stretches forward into the

First of all make sure that your dog feels comfortable in the starting position, and then slowly pull him forwards.

Equipment: FitPaws pimpled peanut ball.

standing position. Make sure that his neck is in a straight line with the rest of his spine.

7. The dog should hold this stretched position for at least 15 seconds. Give him a generous reward for this!

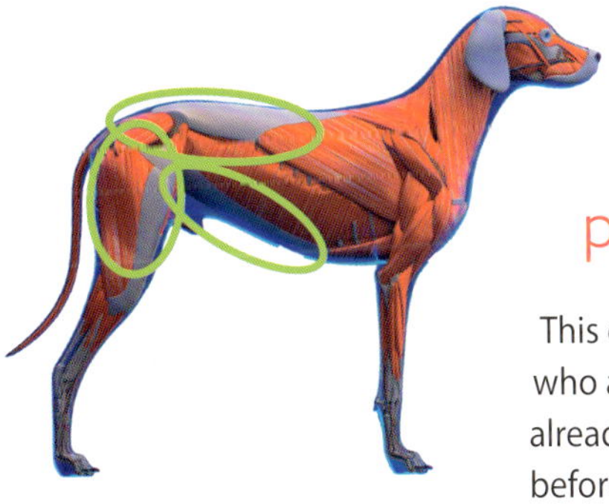

Exercise Band 1

6. Kick-back on the peanut ball (Professional)

++ CO-ORD

+++ STRENGTH

This exercise is only recommended for sports dogs who are very fit. It is essential that the dog has already mastered walking backwards on to a target before attempting this exercise.

1. Place a mat in front of a chair so that the distance between mat and chair is approximately the same as the dog's body length. If the dog were to stand directly in front of you, his back feet would be on the mat.

2. With a clicker and treats at the ready, ask the dog to walk back. Watch him closely and mark the moment when he steps back on the mat.

3. Repeat this at least 10 times.

4. Now increase the size of the target mat, or replace it with an elevation. Proceed in small steps. The rest of the exercise remains the same: click the dog the moment he steps on to the target with both back feet.

5. For this exercise, the aim is not to build up distance, but to focus on targeted, intensive steps backwards.

6. Keep increasing the height of the target. You can also work with a slanting board and gradually make it steeper by using cushions, or something comparable. However, make absolutely sure that the dog cannot injure himself.

7. At this stage, some dogs will develop a powerful kick-back on their own. If so, you can transfer this behaviour directly to the peanut ball. As the peanut is springy, it will increase the effect of the kick-back. Other dogs are more likely to walk back to the target, and will not develop a real kick-back. If this is the case, it is a good idea to work with a cavaletti pole, which you place in front of the target to influence the stride length. The distance to the target is highly individual so it is a matter of experimenting. You will know the exercise is being performed correctly when the dog kicks back, like a young horse, in the direction of the peanut ball and lands his back feet on it.

The dog is sent backwards on to the peanut ball.

Equipment: FitPaws peanut ball.

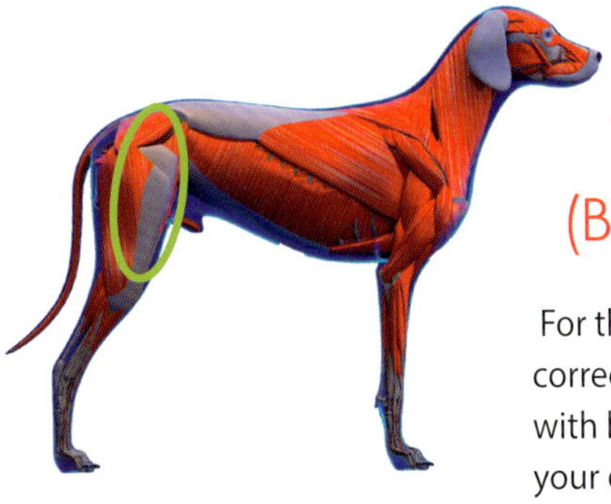

Exercise Band 2

1. Lifting the buttocks (Basic)

+ CO-ORD

++ STRENGTH

For this exercise, it is important that your dog is correctly positioned, i.e. lying in place, like a sphinx, with his hind legs angled and close to his body. If your dog has problems with this, you should have him checked over. If there is no physical issue, you can train him to lie in the correct position by using a platform.

1. Use some mid to low-value treats and refresh the hand touch at least three times. Alternatively, work with a Post-it and the nose target, which will be less likely to hurt your back!
2. Position the dog in the correct lying down position.
3. Offer the target directly in front of him so that he only has to stretch his nose a little forward at first. Reward this first, barely noticeable, movement.
4. Now try to find the perfect distance from the dog's nose. The aim is for him to remain lying down, not moving forward, and only lifting his buttocks slightly, thus activating the hindquarters.
5. The movement towards the treat tends to tense the front thigh muscles; the movement away from the treat relaxes the muscles and lets the dog sink back into the starting position. The result is a flowing, but highly intensive movement.

The dog should only lift his buttocks.

Equipment: Target stick.

Tip: You can also work with an object, such as a child's toy, which can be used as the target for a nose touch.

CO-ORD +

STRENGTH ++

Exercise Band 2

2. Crawling (Advanced)

1. Ask the dog to go into a down.
2. Attach a Post-it to your foot and have a clicker at the ready.
3. Click for any interest in the Post-it. A paw touch would be ideal, but a nose touch is also fine.
4. Click and reward this interest several times.
5. If the touch is well established, you can start to move slowly backwards, at first only half a step.
6. The dog will try to follow. Click and reward the smallest movement. Make sure you always feed your dog in position, as this will stop him getting up.
7. Continue to do this gradually, until you can move several steps without stopping to reward. Your dog will develop his own unique crawling technique.
8. Introduce a verbal cue.
9. Slowly fade the Post-it by making it smaller and smaller.
10. Vary your position in relation to the dog.

The aim is for the dog to crawl a short distance.

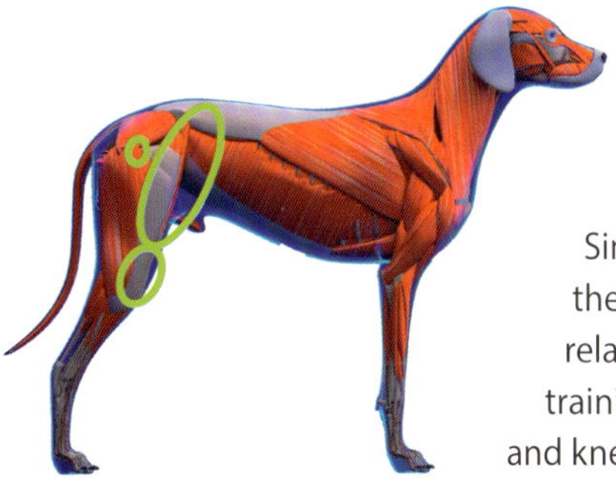

Exercise Band 3

1. Narrow Cavaletti (Basic)

+++ FLEX

++ CO-ORD

+ STRENGTH

Since this exercise is about strength training, the poles of the cavaletti should be narrow and relatively high. The ideal height for strength training with cavaletti is between the dog's hock and knee height, or just above. The distance is chosen so that the dog has to work a little harder without having to take steps in-between. Ideally, you should work with at least five cavaletti poles. As a visual aid, you can set up the exercise next to a wall.

Unless the dog is very experienced, it is essential that you work with him on a leash and harness, especially if this exercise is new to him. Don't let him fall into agility mode and attempt to leap over the poles. The structure of cavaletti training can be adapted to the character of your dog in many different ways. Interestingly, low cavaletti, which are designed to lengthen the stride, are more likely to encourage the dog to speed.

Training options:

- *For fast dogs: First train running from one ground target to another – the dog must position his front feet on each target, in turn. The targets should start close together, and the distance should be gradually extended. Then place one pole between them, then two, and so on.*
- *For food obsessed dogs: Calmly lead the dog over the cavaletti using a hand target. A food reward is only given at the end of the set.*
- *For clumsy dogs: Hide treats between each cavaletti pole. Use large treats so the dog has to chew for a long time.*

For more cavaletti training ideas, see page 204.

For this exercise, the dog should step upwards and away from himself.

Equipment:4-6 Cavaletti sets.

FLEX ++

CO-ORD ++

STRENGTH +++

TACTILE +

Exercise Band 4

1. Climbing stairs (Basic)

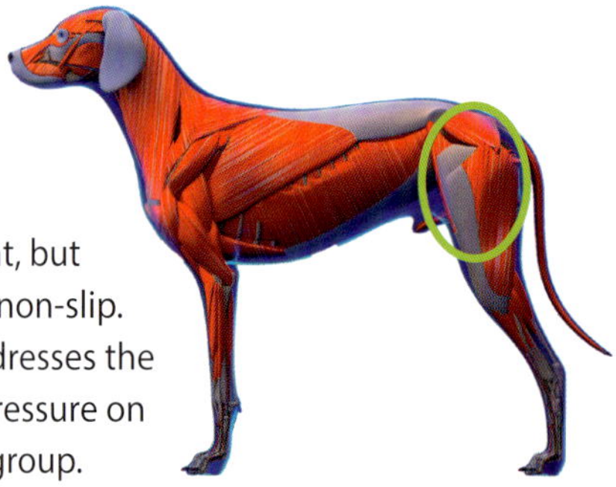

Climbing stairs can be simulated with equipment, but you can also use normal stairs, which should be non-slip. Bear in mind, it is only climbing upstairs that addresses the front thigh muscles; climbing downstairs puts pressure on the forehand, which is not the targeted muscle group.

1. Lead the dog to the steps with your target hand.
2. Reward climbing the stairs. Make sure you feed your dog in position, i.e. one back foot is on a lower step and the other on a higher step. Remember, the slower the dog goes, the more effective the exercise.
3. At the top of the stairs, your training begins! Lift your dog down and let him climb the stairs again. Of course, you can also let your dog climb back down the stairs, but this exercise is more of a full-body workout.

Lead the dog up the steps slowly, one at a time.

Equipment: Stair-like construction.

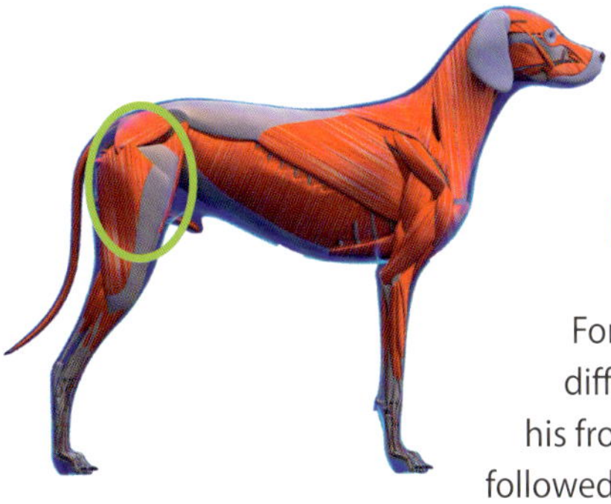

Exercise Band 4

2. Stepping up with hind legs/wobbly (Advanced)

++ FLEX

++ CO-ORD

+++ STRENGTH

+ TACTILE

For this exercise, you need three targets of different heights. The dog should stand with his front feet on a stool or something comparable, followed by a medium-high target in the middle, and a low target at the far end. The aim of the exercise is for the dog to step with his hind legs from the medium-high target to the low target; this is similar to step aerobics. Ideally, you should work on this exercise in front of a mirror or with an assistant.

1. First, position the stool.
2. Lead the dog to the stool and, with your clicker and rewards at the ready, click for initial interest. You can then, gradually, shape him to put his front paws on the stool. Reward this several times.
3. Now give the dog a short break so that you can set up the rest of the equipment. Choose the distance between the middle and lowest targets so that the dog is positioned exactly between them when he stands relaxed with his front feet on the stool. This will help him understand the exercise
4. Before you set the dog to work again, make sure you are in a position to observe him closely, and that you will be able to reward the first step in the new set-up.
5. This is because the dog now has to walk over the added equipment in order to

Now you are asking the dog to use his hind legs to step on to a wobbly surface.

Equipment: Ikea stool (tall), FitPaws FitBone, TOGU Aero-Step One.

replicate his previous behaviour; his progress should, therefore be marked and rewarded directly. If he tries to bypass the other targets, he will not receive a click. If necessary, help him with a hand target. Alternatively, you can build an alley with cavaletti poles, which will act as a visual aid.

6. Now things are getting exciting. The aim of this exercise is for the dog to swing back and forth, with one hind leg at a time, from the middle target to the lowest target. The dog will decide which leg you click first, because the first one he offers is the one you will reward. You can shape the back and forth movement, or you can offer a hand touch or a chin target at the front so that he stretches forward a little and pulls one leg forward. This is move is fleeting – a matter of seconds – which is why working with a mirror, or assistant, is essential.

7. If the dog is performing the exercise on one side, you can introduce a verbal cue, and then train the other side. Some dogs also offer both sides directly, which makes your work far easier!

Tip: If your dog doesn't offer any movement with his hind legs, you can place a cavaletti pole between the targets to give him a clearer idea. In this case, it is easier to work on the complete set-up rather than introducing the stool in the preliminary stages.

CO-ORD +

STRENGTH +++

TACTILE +

Exercise Band 5

1. Lunge forward (Basic)

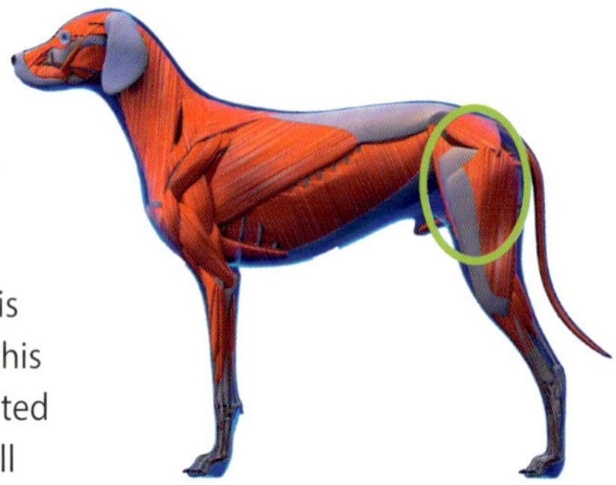

The aim of this exercise is for the dog to thrust forward and shift his weight. You can achieve this with the help of a hand touch or chin target. In this exercise, you can, ideally, train one leg in a targeted manner by not placing it on an elevation and still putting weight on it. Recruiting an assistant, or using a mirror, is recommended for this set-up.

1. For this exercise, you will need a stool and two targets for the dog's back feet. Ideally, these should be elevations so he has a visual aid. Before you start, the dog should already be familiar with placing his front feet on a stool.

2. Select the targets for the back feet, and then choose one of the following two training options:

Option 1 – Start with one target:

Set up a target and, using a hand target, lead the dog over the target from behind so that he steps on it with one back foot. If this does not work, use an alley as an aid. Make sure you remember which foot he used to step on the target. Feed him in position, i.e. click and reward him for standing still on the target. After approximately 10 clicks, which the dog earns for maintaining his position, throw the food to the back. This means he has to leave his station to get his reward, and then return to the target to earn his next click.

Tip: if your dog avoids the target with his back foot, replace it with something larger and reward with extremely high quality food, when the dog steps on the new, larger target.

Repeat this automatic target-finding process several times. If this works reliably, insert the second target. Make sure that your dog can stand on it comfortably when he is shifting his weight forward, i.e. with one foot on the target, and one behind it. If necessary, you can help by 'pulling' the dog forward with a hand touch. Keep practising until your dog is standing on both targets, in the desired position, before introducing the stool for the forehand. Using a front elevation puts more weight on the hindquarters and, therefore, more power is needed to remain in the lunge position. As the dog is already familiar with the stool, putting the exercise together should be relatively straightforward.

Option 2 – start with both targets:

Proceed in the same way as above, except that you already have both targets at an appropriate distance from each other. Depending on the dog, this procedure may be easier to implement. Large targets make training easier.

The dog should hold his position and not change legs.

Equipment: Ikea stool (large), TOGU Aero Step One.

CO-ORD +++

STRENGTH +++

Exercise Band 6

1. Weight shift on a wobble board (Basic)

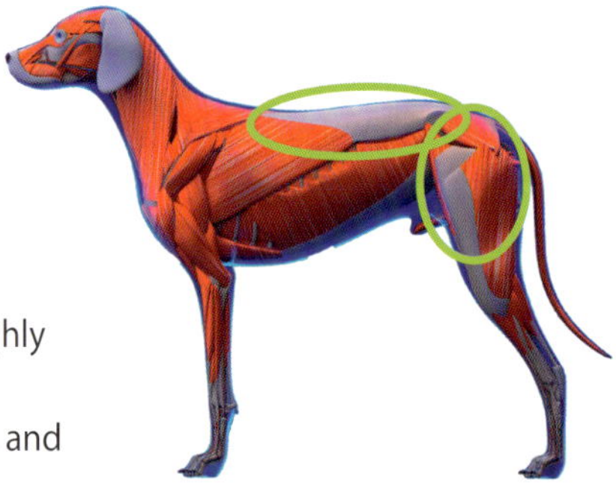

This exercise, although it looks very simple, is highly effective. You need a wobble board, which is of sufficient size for the dog to stand on in comfort, and off you go!

1. First of all it is very important that you stabilise the wobble board. You can use cushions or balance equipment, which you should place underneath all four sides of the rocking board to prevent excessive wobbling.
2. Take time to introduce the dog to the wobble board; he needs to be relaxed in order for the training to be successful.
3. Initially, ask the dog to sit in front of the wobble board. It is important to establish this as a basic ritual to ensure he doesn't climb on to it, on his own initiative, in the future. Use your hand to stabilise the board and allow him to stand on it.
4. If the dog feels comfortable on the stabilised wobble board, and you have clicked and rewarded him for standing on it several times, you can remove the aids you have used to stabilise it. First remove the stabilisers that are in front, and behind, the dog. Secure the wobble board with your hand until he has got on it again. Now click and reward for the new set-up, and for staying on the board while you wobble it.
5. Proceed in the same way, removing the stabilisers on the right and left sides.

For this exercise the dog stands still and shifts his weight from behind to the front and back.

Equipment: FitPaws wobble board.

6. Now the dog is familiar with the wobble board, which wobbles front and back, and from side to side, it's time to train the exercise. The aim is for the dog to shift his weight without moving his feet, which he can do by initiating a rocking movement. You can trigger this by using a chin or a nose target, depending on what best suits the dog. Choose the option which is most likely to be successful, and proceed slowly. Click for every slight weight shift, or seesaw movement, of the board and reward the dog generously.

7. Over time, the dog will develop a real love for this exercise, and will be able to control the degree of wobble on his own. You can also try wobbling the board from side-to-side to activate the long back extension diagonally

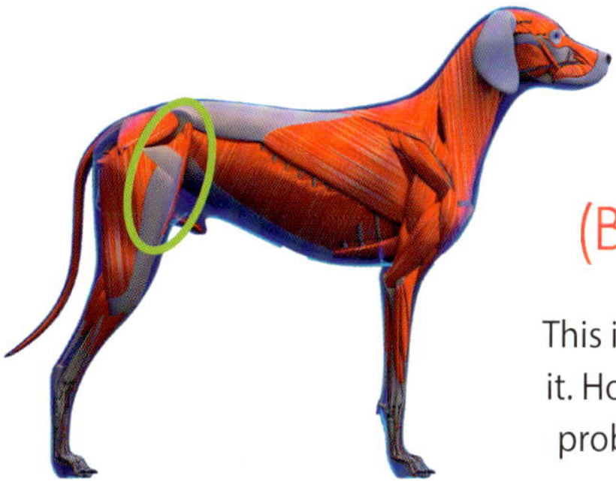

Exercise Band 7

1. Bounce to a hand touch (Basic)

++ CO-ORD

+++ STRENGTH

This is a fun exercise and dogs enjoy performing it. However, it not suitable for dogs with back problems.

1. Ask the dog to sit.
2. Now offer him a hand target, just above the tip of his nose, so that he has to stretch straight up to reach it. Click and reward.
3. Repeat this several times.
4. Now slowly move the hand target higher so the dog learns to push off with his forehand to reach it. Make sure he doesn't stand up; his bottom should stay on the ground and he should not turn his hind legs outwards. Click and reward the dog for keeping his legs in a parallel position. If he stands up, or slides out with his hind legs, interrupt the trick briefly and ask the dog to sit up straight again. Make sure he doesn't try to use you as a support; you can avoid this by keeping an appropriate distance from him.
5. Proceed with caution. This exercise is very strenuous and the dog must first find his balance. Give him time to do this.

Be careful: If the hand target is held too high, the dog cannot keep his bottom on the ground.

CO-ORD +++

STRENGTH +++

TACTILE +

Exercise Band 7

2. Bounce to a hand touch/wobbly (Advanced)

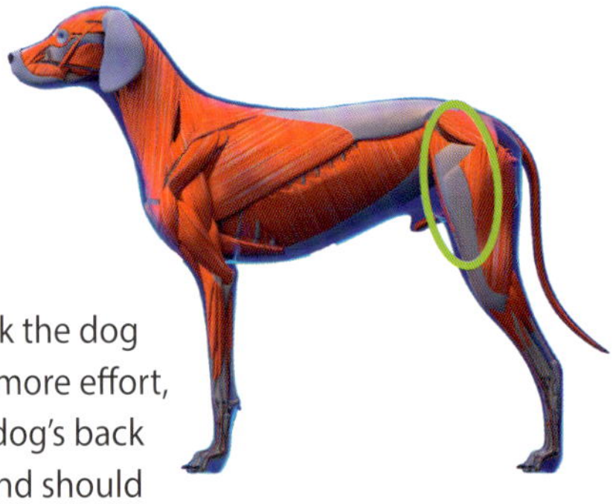

To intensify the basic exercise (see page 111), ask the dog to work on a wobbly surface. This will take a lot more effort, and it is important that it is done correctly. The dog's back feet should be parallel to his body at all times, and should not turn out. If the feet turn out, it is an evasive position and will not have the desired effect on the front thigh muscle.

Transfer the exercise to a wobbly surface.

Equipment: TOGU Dynair Extreme.

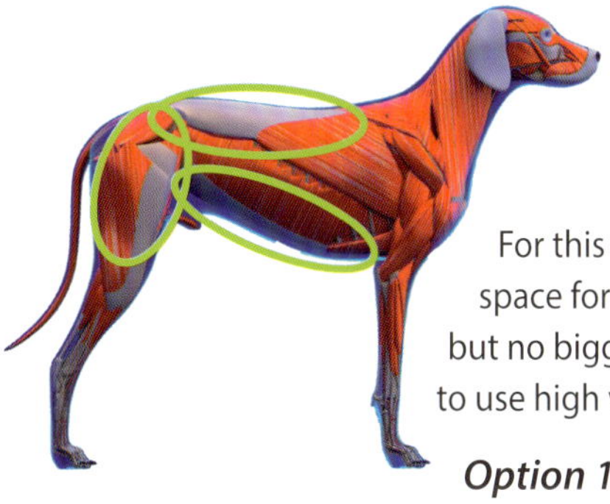

Exercise Band 8

1. Mountain goat (Basic)

++ CO-ORD

+++ STRENGTH

For this exercise you need a stool which has sufficient space for the dog to stand with his feet close together – but no bigger – just like a mountain goat! It's a good idea to use high value treats when you are working on this.

Option 1 – Set-up over two elevations

This set-up is suitable for cautious dogs and also for those who are easily excited.

1. Place the stool, which you intend to use for the finished exercise, behind another elevation. You can work with another stool, but you could also use your sofa, for example.
2. Lead the dog from behind on to the first elevation and reward him, in small steps, for climbing on to it. Use low-value kibble for this. The dog may stand on it with all four feet, or he may place his front feet on the second elevation – the stool. If he uses his front feet, reward him with high-value treats.
3. Now guide the dog, either by luring, or by using a hand target, so that he places all four feet on the stool.
4. Make sure that the dog is standing with his feet close together; his back should be slightly rounded, or straight. If he overstretches his head, make sure he does not hollow his back.

Option 2 – Stepping directly on to an elevation

With experienced dogs, you can start right away with the desired elevation.

1. Arm yourself with clicker and treats, and lead the dog to the elevation from behind. Now the climbing begins: encourage the dog to climb the hill. Reward even the smallest steps in the right direction.

Here, you can see the mountain goat position, with focus on the front thigh area.

Equipment: Ikea stool (large).

2. When the dog is standing securely on the elevation, use a hand touch to change his head position. Make sure that he does not hollow his back. The ideal posture is when the neck is at the same level as the spine, drawing a line between the cervical and lumbar spine. In order to shift weight backwards momentarily, you can hold out your hand so the dog is looking up, and his thigh muscles are more active.

CO-ORD +++

STRENGTH +++

Exercise Band 8

2. Mountain goat/wobbly (Advanced)

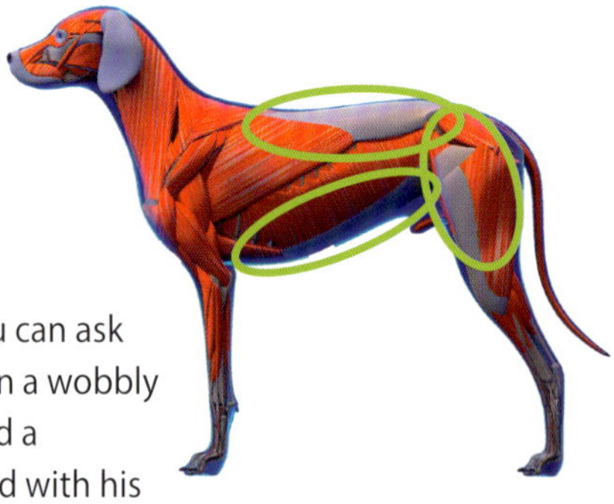

To progress the basic exercise (see page 113) you can ask the dog to go into the mountain goat position on a wobbly surface. Again, you will need to find a surface that allows the dog to stand with his front and back feet close together. When the dog is standing on it in a relaxed position, you can start varying his head position.

Initially stabilise the therapy gyro by placing small balance discs underneath it.

Equipment: Therapy spinning top.

V. EXERCISES FOR THE OUTER THIGH MUSCLES AND HIP EXTENSOR

Outer thigh muscles and hip extensor		
Category:		***Hind limbs***
Training goal:		*Strengthening, co-ordination and greater range of movement*
Recommended:		*For dogs with under-developed thigh musculature, for stabilisation during sporting activities and in the case of health deficits (e.g. HD)*
Muscular function:		• *The biceps femoris muscle (two-headed thigh muscle) with its two parts takes over an important function in the stance phase/gait (30-50 per cent), sideways pulling, stretching the hip, bending the knee and stretching the ankle* • *The gluteal muscles (croup muscles) are primarily responsible for stretching the hips, for forward lifting and participate in abduction (sideways pulling). In addition, the adductors and the so-called 'hamstrings' are brought into use* • *The outer thigh muscles are most affected in terms of muscular atrophy* • *If the thigh muscles (outside) are atrophied, the adductors (inside) are usually under greater tension, so the lateral exercises are perfect for this.*
Outer thigh muscles ①	CO-ORD ++ FLEX ++ STRENGTH +	***1.) Stepping sideways over a pole (Basic)*** ▸▸ *Builds up the outer thigh muscles and 'forces' the adductors (inside) to give way.*
	CO-ORD +++ FLEX + STRENGTH ++	***2.) Stepping sideways over an elevated pole (Advanced)*** ▸▸ *Builds up the outer thigh muscles and 'forces' the adductors (inside) to give way.*

Outer thigh muscles ①	CO-ORD +++ FLEX +++ STRENGTH ++	***3.) Stepping sideways over a series of poles (Professional)*** ▸▸ *Builds up the outer thigh muscles and "forces" the adductors (inside) to give way, additionally bending the knee*
Outer thigh muscles ②	CO-ORD ++ FLEX ++ STRENGTH ++	***1.) Elephant trick (Basic)*** ▸▸ *Development of outer thigh muscles and increasing load on the hind limbs*
	CO-ORD +++ FLEX +++ STRENGTH ++	***2.) Elephant trick stepping over one or more poles (Advanced)*** ▸▸ *Additional co-ordinative component to the basic exercise*
	CO-ORD +++ FLEX +++ STRENGTH ++	***3.) Elephant trick variations (Advanced)*** ▸▸ *Additional strengthening of the hindquarters*
Outer thigh muscles ③	CO-ORD +++ STRENGTH ++ FLEX ++	***1.) Sloping board position (Basic)*** ▸▸ *Force is absorbed on the stabilising legs; the legs positioned on the slanting board activate the outer thigh muscles, and the adductors on the inside are actively stretched*
	CO-ORD +++ STRENGTH +++ FLEX ++	***2.) Sloping board position at a distance (Advanced)*** ▸▸ *The same effect as the basic position but involves more strength*

Outer thigh muscles (4) **Isometric exercises for the thigh muscles**	CO-ORD ++ STRENGTH ++	**1.) Isometry: Lifting front legs (Basic)** ▸▸ Strength exercise without joint movement, focus on outer thigh muscles
	STRENGTH ++ CO-ORD + TACTILE +	**2.) Isometry: Thigh pressure (Basic)** ▸▸ Strength exercise without joint movement, focus on the outer thigh muscles
	STRENGTH +++ CO-ORD ++ TACTILE +	**3.) Isometry: Lifting back legs (Basic)** ▸▸ Strength exercise without joint movement, focus on the outer thigh muscles
	STRENGTH +++ CO-ORD ++ TACTILE +	**4.) Isometry: Wobbly surface (Advanced)** ▸▸ Strength exercise without joint movement, focus on outer thigh muscles
Outer thigh muscles (5)	CO-ORD ++ FLEX ++ STRENGTH +	**1.) Side stepping with front and rear limbs stationed (Basic)** ▸▸ Similar effect to the elephant trick, except the legs are not abducted as much (small steps) and all four limbs are equally loaded

FLEX ++

CO-ORD ++

STRENGTH +

Exercise Band 1

1. Stepping sideways over a pole (Basic)

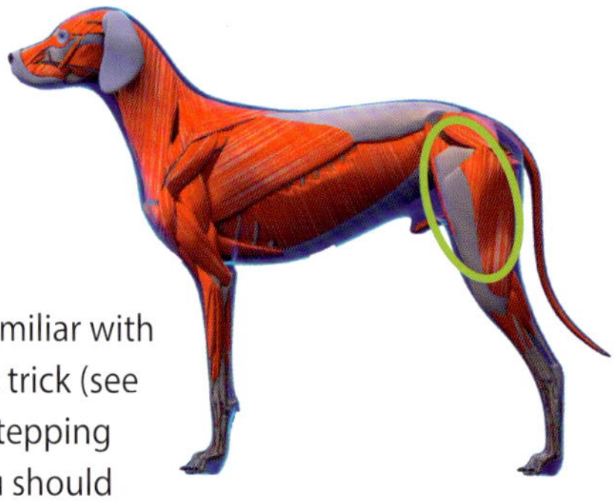

For this exercise, it helps if your dog is already familiar with stepping sideways or, at least, with the elephant trick (see page 35), as he will understand the concept of stepping sideways over a pole more quickly. Basically, you should pay particular attention to your dog's hindquarter action in this exercise; the forehand will then follow as if by magic.

Option 1: via elephant trick

First, refresh the elephant trick. Transfer this to the floor by selecting lower and lower elevations from your original elevation until, finally, you are working with a target mat, or similar.

1. Position yourself in front of the dog and practise one or two sideways steps. Some dogs have no problem with moving away from the floor target and will, willingly, follow you. If this is the case, reward generously. Others remain steadfast on the target. If this happens, try adding a target, i.e. place another mat next to the original mat so that the dog can step over to the next one.
2. If the dog is stepping sideways with confidence, you are ready to place a pole next to the target so that he has to step over it with forehand and hindquarters when following you in the front position. You can use your feet to stop the pole rolling away. Reward generously.
3. If you have a 'wait-and-see' dog, you may be able to help him with a hand target. However, it is important to click the hindquarters crossing over, as otherwise your dog's focus will be in front, and he will not think about his backend.
4. Generalise the exercise with different poles, pole heights, and also without floor target(s).

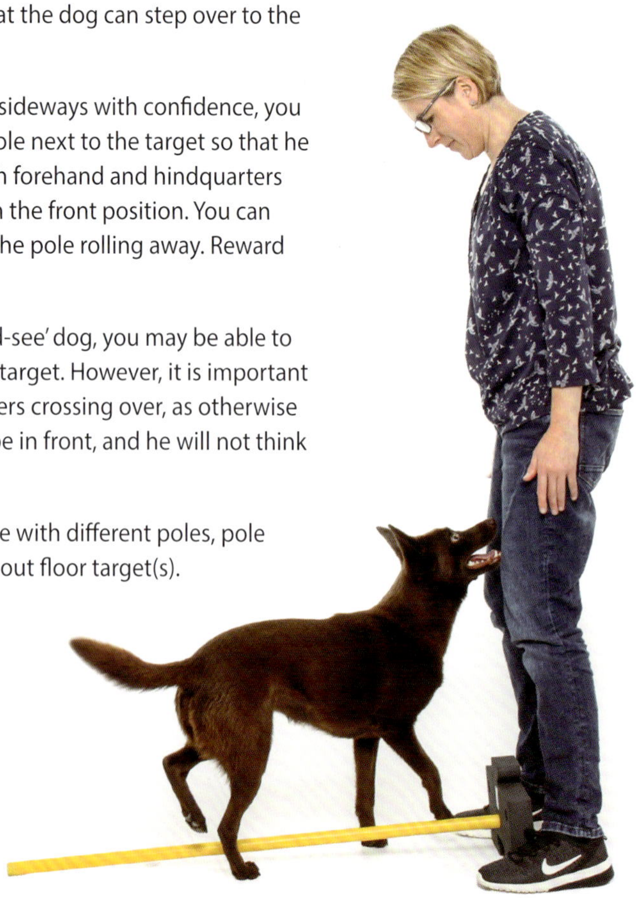

Initially elevate the front of the pole only, until the dog becomes more experienced.

Equipment: Dog-stepper with cavaletti.

Variant 2: via slanting board

1. If the dog is already familiar with moving sideways on to a sloping board (see page 124), refresh it a little and reward the behaviour.
2. Now place a cavaletti pole in front of the slanting board in such a way that there is room to step over it first.
3. Ask the dog for the slanting board position but, as he steps over the pole, click and reward. It is important to click as he puts his hind leg back on the floor, rather than when he is lifting it to step over the pole, so that he repeats this behaviour.
4. Now fade out the slanting board by moving it further and further away from the cavaletti pole.
5. You can now introduce a verbal cue.

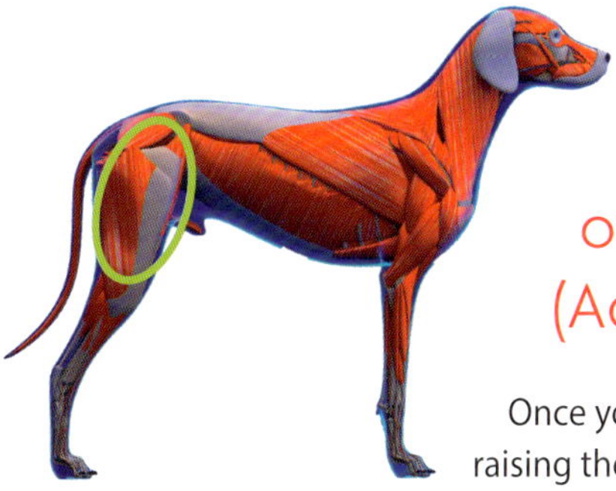

Exercise Band 1

2. Stepping sideways over an elevated pole (Advanced)

\+ FLEX

+++ CO-ORD

++ STRENGTH

Once your dog has understood the basic exercise, raising the pole should not pose a problem. However, you should proceed gradually until the pole is at maximum, hock height. Remember that this exercise is strenuous, which means that you should finish a session before the dog shows signs of fatigue, such as becoming hesitant or clumsy.

For this exercise, adjust the pole to maximum hock height.

Equipment: Cavaletti set.

FLEX +++

CO-ORD +++

STRENGTH ++

Exercise Band 1

3. Stepping sideways over a series of poles (Professional)

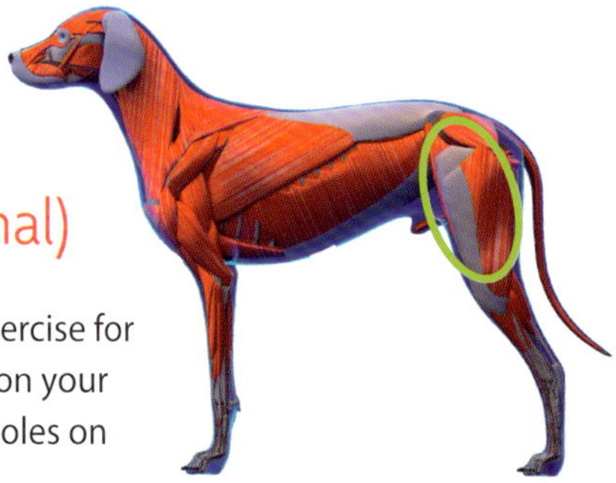

Stepping sideways over a series of poles is an exercise for professionals only. If your main focus is to work on your dog's co-ordination, it is sufficient to place the poles on the floor, or elevate them slightly.

For strength training, the poles will need to be higher but, again, no higher than wrist height. Do not over strain your dog!

Stepping over a series of poles is an extremely challenging exercise.

Equipment: Several cavaletti sets.

Exercise Band 2

1. Elephant trick (Basic)

++ FLEX

++ CO-ORD

++ STRENGTH

1. You will need an elevation for the dog's front feet. You can start with a stool, which will make it easier, or you can opt to use a balance cushion from the start.
2. Get ready with a clicker and treats, and click the dog's initial interest in the stool/balance cushion. Keep working until he gets the idea of placing his front feet on it. Reward generously so that he remains on the stool. If he is reluctant to place his front feet, you can help with a hand target.
3. When the dog is staying in position on the stool/balance cushion, position yourself directly in front of him and offer a hand target at muzzle level. Repeat this at least five times.
4. Now you can start to circle slowly around the stool/balance cushion. Take tiny steps at first, so that the dog only has to stretch his head to the side. Click and reward.
5. Now move a little further away from the dog, but in the same direction as before. Position the hand target centrally, in front of your body. Now the dog must turn slightly. Observe closely and make sure you click as his hindquarters move. As soon as he moves sideways, click and reward. The dog should now be standing straight in front of you again, so you can work on a number of repetitions, clicking and rewarding each time he moves to the side.
6. If the dog has understood the exercise, and is moving sideways in one direction, you are ready to start work on the other side, following steps 3 and 4.
7. Once the dog is reliably offering the behaviour on both sides, you can introduce a verbal cue.

Tip: If the dog's hindquarters are not moving, help him by moving your target hand towards his shoulder. He should follow this movement and will, involuntarily, move his hind legs. Click and reward this several times and practise in both directions.

Starting position with front orientation.

The dog rotates while keeping his front feet on a balance cushion.

Equipment: Balance cushion.

FLEX +++

CO-ORD +++

STRENGTH ++

Exercise Band 2

2. Elephant trick stepping over one or more poles (Advanced)

1. Refresh the basic *elephant* trick (see page 35).
2. Place a cavaletti pole next to the elevation and lead the dog from behind, over the pole, so that his back feet are on either side of it.
3. Observe closely to work out which side is easier for your dog. Click and reward for stepping over the pole.
4. If the dog is fluent in one direction, start work on the other direction.
5. Once the dog understands the exercise, and is offering the behaviour in both directions, you can add more poles.

For these exercises cavaletti poles are positioned between the TOGU Dynair ball iron and TOGU Brazil.

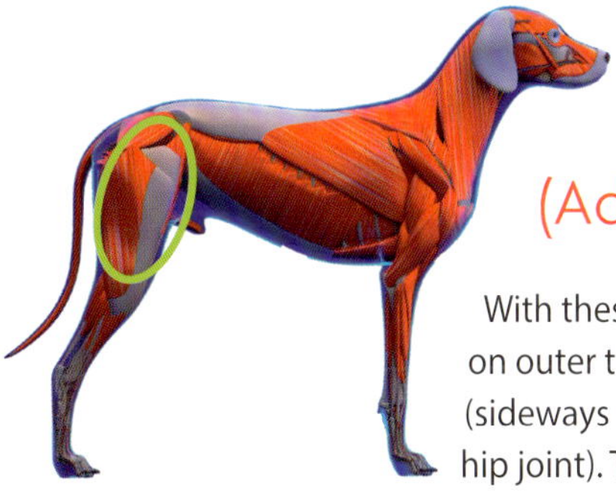

Exercise Band 2

3. Elephant trick variations (Advanced)

+++ FLEX

+++ CO-ORD

++ STRENGTH

With these elephant trick combinations, you are focusing on outer thigh muscles, such as the biceps femoris muscle (sideways stepping) and the gluteal muscles (extensor hip joint). These exercises are also suitable for dogs with stiff/restricted movement of the hind limbs. They should be encouraged to shift their weight to the rear, i.e. relieving weight on the front limbs.

Option 1: Balance cushion in the middle, yoga blocks/paw pods in a circle around it

With this combination, you are addressing dexterity and sensory motor skills. The training goal is for the dog to 'find' all objects around the balance cushion with his back feet while keeping his front feet in the middle of the balance cushion. This should be done cleanly, without pushing the objects around.

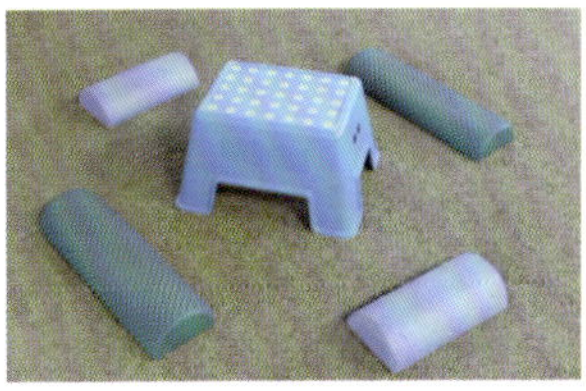

Option 2: Elevation in the middle, yoga blocks/paw pods around it

A similar combination but, this time, the dog's front feet are positioned on an elevation, such as a stool. The increased elevation of the forehand means the dog has to work harder with his hindquarters so that he has sufficient strength to move sideways.

Option 3: Small elevation and cavaletti poles

Here, the dog is standing on a small, yet stable, elevation with his front feet, while he steps over the cavaletti poles with his back feet.

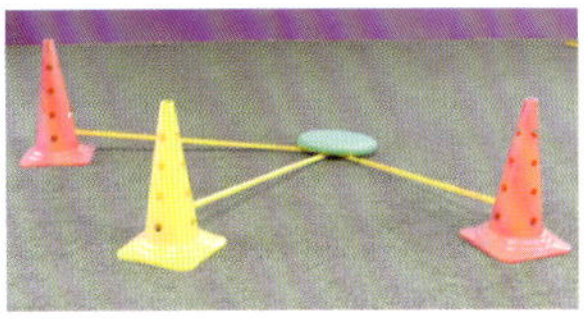

Option 4: Unstable at the front, pole stepping at the back

The dog must position his front feet on a wobbly elevation, such as a balance cushion, and move sideways to step over the poles. He needs to balance at the front and step a little higher over the poles with his back legs.

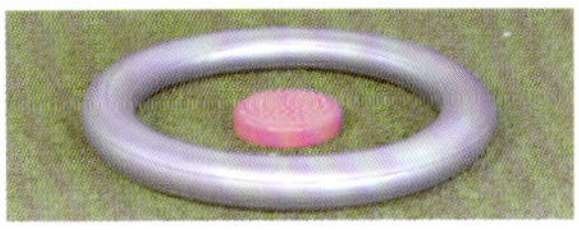

Option 5: Around the ring

This is a great exercise for small dogs. The outer ring, which is inflatable, also promotes co-ordination.

FLEX ++

CO-ORD +++

STRENGTH ++

Exercise Band 3

1. Sloping board position (Basic)

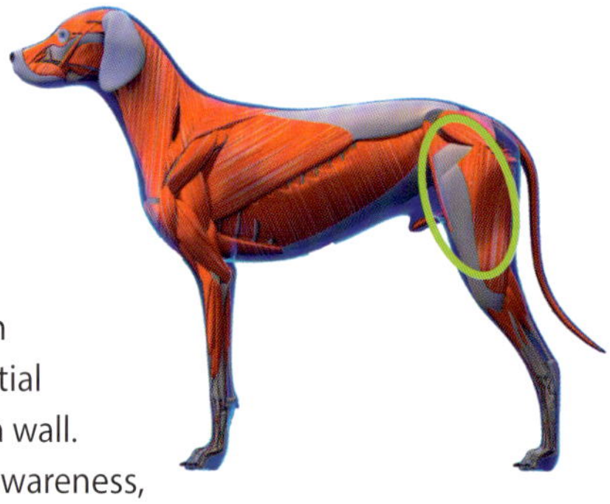

For this exercise you need a sloping board which corresponds to the length of your dog. In the initial stages of training, it should be lined up against a wall. To be successful, the dog needs good rear-end awareness, which he can acquire by working on the elephant trick (see page 35). If your dog is already at an advanced level with his sideways stepping, you can use the sloping board on its own, without the added support of a wall. If your dog needs more help, proceed as follows:

1. Place the sloping board directly against a wall, at an angle of approximately 15 degrees. This means the dog will not confuse it with a back or forehand target, or with his platform training, but receives a clearly defined signal telling him "this is another exercise". Make sure the sloping board is secure and will not move.
2. Lead the dog from behind on to the sloping board and reward him for standing on it. Initially, this will probably be with all four feet. Then, throw the treat behind him so that he has to leave the board to get his treat, and then return to it in order to earn his next reward
3. Some dogs will, by chance, stand with two feet (front and back) on the sloping board and the other two on the floor. If this happens, reward the dog generously.
4. From now on, if the dog adopts the correct position, mark it by feeding high-value treats. If he has all four feet on the sloping board, you can still reward, but use low-value treats.
5. Now it becomes a test of your throwing skills. First, throw the treats backwards in a straight line; this would be the equivalent of 12 o'clock on clock-face. Gradually change your aim until you are throwing towards 3 o'clock.
6. Initially, the dog will run forwards on to the sloping board. That doesn't matter; the only criterion that changes is that you only reward him when he is standing with two feet on the sloping board and two feet on the ground. If he stands with all four on the board, encourage him to try again.

The wedge cushion can be stabilised against a wall until the dog fully understands the exercise.

Equipment: Wedge cushion.

7. If the dog understands what is required, you can start to gradually increase the height of the board. Be careful not overdo it, as the dog needs to be comfortable. A maximum spread of 70 degrees is recommended.

8. Now introduce a verbal cue for finding and holding the position. For maximum benefit, the dog should hold the position for 10-15 seconds. If necessary, you can use the chin target to help him maintain his position as this helps him to understand that it's only a matter of standing still.

Exercise Band 3

2. Sloping board at a distance (Advanced)

++ FLEX

+++ CO-ORD

+++ STRENGTH

If your dog is already familiar with the sloping board, you can now add some distance. This is a greater challenge as he has to take more sideways steps to get to the board.

1. To do this, start by asking him to stand in front of you. Now stand next to the board and ask the dog to step sideways on to it. Reward, and repeat several times.

2. Now move the dog just a few centimetres to the side. If he can find the sloping board from there without any problem, click and reward him immediately with a high value treat.

3. Now you can gradually increase the distance. The dog will work out what he has to do, and he may even start 'searching' for the board. This shows he has a full understanding of the exercise and it also increases the training effect.

The wedge cushion can be stabilised against a wall until the dog fully understands the exercise.

Equipment: Wedge cushion.

CO-ORD ++

STRENGTH ++

Isometric exercises for the outer thigh muscles

Exercise Band 4

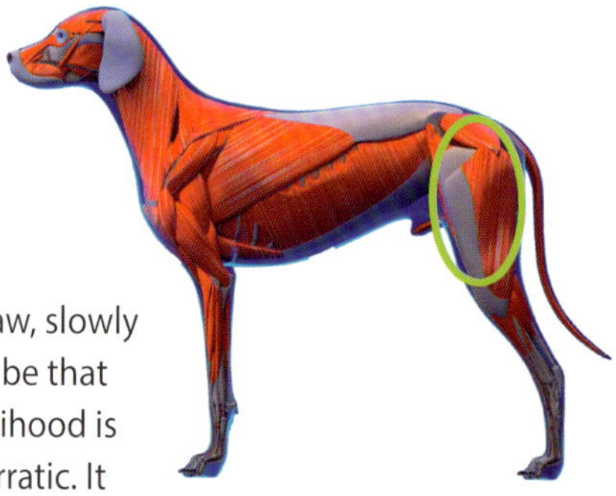

1. Isometry: Lifting front legs (Basic)

For this exercise the dog needs to lift his front paw, slowly and carefully, and then hold the position. It may be that your dog knows how to 'give a paw', but the likelihood is that the movement he offers will be quick and erratic. It therefore makes sense to touch the leg and then move it gently upwards, supporting the elbow with your other hand from below. This will cause the dog to weight shift, thus activating his hindquarters as they are required to bear more weight.

You can alternate front legs for this exercise, asking the dog to hold position for approximately 10 seconds per side.

If the dog has a muscle deficit on one side, you can focus on additional training for this side. If he attempts to sit down while you are doing this, he will need a check-up to establish whether he is in pain.

When the front leg is lifted, the dog has to put more weight on his diagonal back leg.

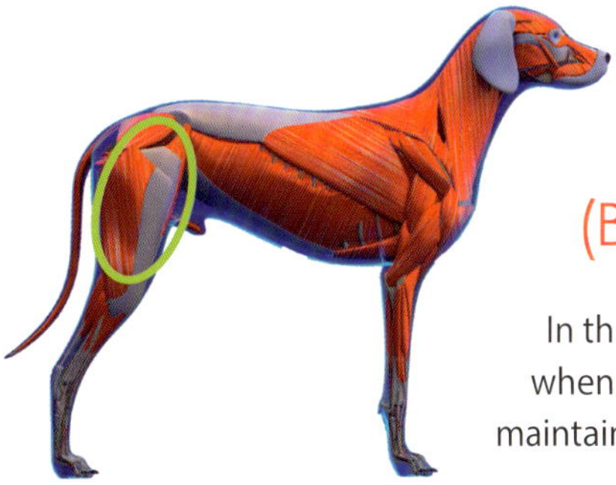

Exercise Band 4

2. Isometry: Thigh pressure (Basic)

+ CO-ORD

++ STRENGTH

+ TACTILE

In this exercise, you apply pressure to the dog's thigh when he is standing; he will have to counteract this to maintain his balance.

1. Position the dog in a stand in front, or beside you. It helps if you place him on a platform, such as a dog-stepper. This gives him security and signals that all he needs to do is stand still.
2. Now proceed slowly, moving your hand near the dog's hip without touching it. Click and reward.
3. If the dog is standing still, you can now make physical contact and touch him gently. Click the moment you touch him, and reward him for holding his position. Don't worry if the dog turns his head at this early stage of training. However it should be avoided in the finished exercise as it has the effect of shifting the dog's weight. If it is an issue, work with a "look" target or enlist a helper.
4. Now intensify your contact so that your hand is resting centrally on his thigh. Click and reward for small, intermediate steps. A high reward rate at this stage pays dividends in the long term.
5. Gradually increase the pressure and mark the moment when the dog reacts with counter-pressure. You can slowly increase and prolong this pressure.
6. You can now introduce a verbal cue, e.g: "Iso". In time, the dog will understand that this means: "when you apply pressure, I will respond with counter pressure".
7. Now you can also train the other side. Remember, the training effect is greatest on the side on which you place your hand. The legs that you are not touching are also kept busy as they are maintaining balance, but they are not working as intensively as the leg that builds up counter-pressure.

The dog should stand parallel – it may help if you recruit a helper for this exercise.

CO-ORD ++

STRENGTH +++

TACTILE +

Exercise Band 4

3. Isometry: Lifting Back legs (Basic)

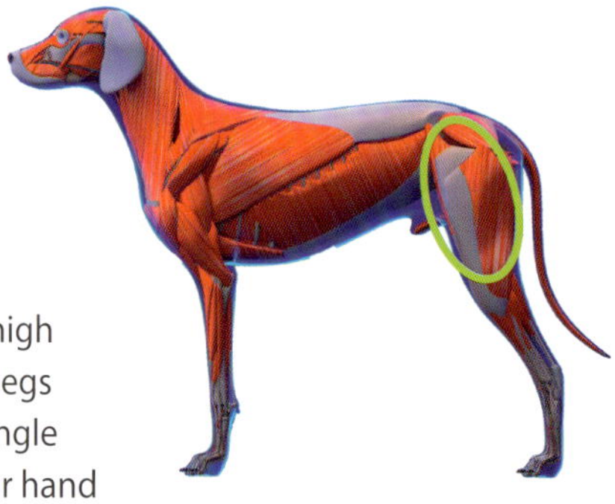

This is another isometric exercise for the outer thigh muscles. All you have to do is lift the dog's back legs while he remains standing. Make sure that the angle is anatomically correct. This involves placing your hand under the dog's foot so that the pressure is applied vertically from top to bottom. Note: your hand should be flat and held parallel to the ground, not tilted.

Tip: Initially, it may help if you support the dog with a chin target. Later, this will become impractical but, when you get to this stage, he will have the confidence to hold the position.

1. Ask the dog to stand on a platform, e.g. the dog-stepper.
2. Touch his hindquarters and click and reward him for remaining in the 'stand'.
3. Now slide your hand down the side of his back leg to the bottom of his foot and leave it there. Click and reward.
4. Slowly lift the foot to a height of 3 to 5 cm, making sure you don't allow the back leg to bend at the knee too much, otherwise the dog cannot build up pressure downwards.
5. Reward with verbal praise, and with some high-quality chewable treats, for staying in position.
6. The aim is for the dog to hold position for 10-15 seconds, allowing you to build up a slight counter-pressure – just enough so that the dog does not feel uncomfortable and tries to swerve away.
7. Proceed in the same way with the other back leg.

This is the perfect height for lifting the back leg – you should not go higher.

Equipment TOGU Ballanza Ballstep.

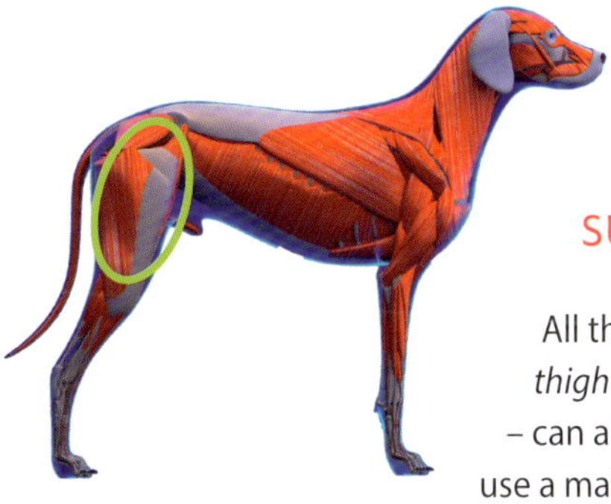

Exercise Band 4

4. Isometry: Wobbly surface (Advanced)

++ CO-ORD

+++ STRENGTH

+ TACTILE

All the isometric exercises – *lifting front legs* (Page 126), *thigh pressure* (Page 127), *lifting back legs* (Page 128) – can also be performed on a wobbly surface. You can use a mattress, a trampoline or special balance equipment, thereby increasing the co-ordinative effect of the isometric training.

If the dog is standing on a wobbly surface when he performs the different isometric exercises, they will have a greater effect.

Equipment TOGU Dynair Extreme.

FLEX ++

CO-ORD ++

STRENGTH +

Exercise Band 5

1. Side-stepping with front and rear limbs stationed (Basic)

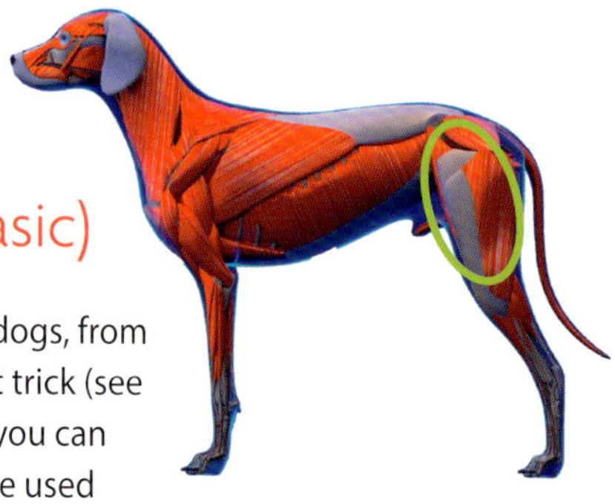

Side-stepping is, in itself, a great exercise for all dogs, from young to old. You can do this using the elephant trick (see page 35) working towards a target on a wall, or you can start with front and rear limbs stationed. We have used the Airex Balance Beam for this exercise, but you could also use rolled up towels, floor targets, or similar.

1. Prepare two floor targets, e.g. seat cushions or two small rubber mats. These are to be used as forehand and backhand targets.
2. Lead the dog from behind over the floor targets, and position him so that he is standing with his front and back feet on the designated targets. Click and reward.
3. Now you need to focus on where you are feeding the dog. The food should be given in such a way that he has to step off the target with one back foot. At this stage, the front feet remain in position on the floor targets. In the following steps, the front feet will, almost automatically, place themselves if the dog has understood the concept of placing his feet on targets.
4. Now click the dog for returning his back foot to the floor target. If he has no idea what you want him to do, help him with a hand target. This may take some time, so repeat this step at least 10 times.
5. After you click the dog, feed him so that he is a little further from the floor targets and has to cover a greater distance to step on to them again. If he manages to do this, reward him generously.
6. Now you can add more targets to create a line.
7. In this scenario, click the dog for stepping on to the targets and feed him when he has stepped down from them so is in the starting position for the next floor targets.
8. Place the targets next to each other and no longer click every single, sideways step.
9. Generalise the exercise with several targets before transferring it to a wobbly surface to intensify it.

Equipment: Airex Balance Beam.

VI. EXERCISES TO RELAX THE BACK MUSCLES

Back muscles: relaxation		
Category:		***Spine / back muscles***
Training goal:		*Relaxation of the back muscles, abdominal muscles, lumbar and iliopsoas muscles*
Recommended:		*For dogs with heavily strained back muscles or with tension in the back, loins and abdominal area (e.g. old scars from castration)*
Muscular function:		• *Back muscles – Bending, stretching, sideways bending, locks the spine, power transmission from hindquarters to forehand* • *Abdominal muscles – Support movement, breathing, bowel movement, urination, labour contractions* • *Lumbar, intestinal muscle – Forward swing of the leg in motion, flexor hip (simultaneously turns knee outwards)*
Back relaxation ①	RELAX ++ FLEX ++ CO-ORD +	***1.) Plank (Basic)*** ▸▸ *Relaxes all the above mentioned muscles and the hip extensors (hold position 10-20 seconds)*
	RELAX +++ FLEX +++ CO-ORD ++	***2.) Plank with lowered head (Advanced)*** ▸▸ *Relaxes all the above mentioned muscles and the hip extensors*

Back relaxation (2)	FLEX +++ RELAX +++	**1.) Lying belly-length over a peanut ball (Basic)** ⏩ Relaxes all the above mentioned muscles
	FLEX +++ RELAX +++	**2.) Lying side-long over a peanut ball (Advanced)** ⏩ Relaxes the lateral back muscles (the side facing upwards). Hold position for at least 10-20 seconds each time
Back relaxation (3)	RELAX +++ FLEX ++ CO-ORD ++	**1.) Stretching (Basic)** ⏩ Make sure the back line is straight. Relax all the above mentioned muscles and the hip extensors. (hold position 10-20 seconds)
Back relaxation (4)	RELAX +++ FLEX +++ CO-ORD +++	**1.) Knee/flank stretching (Basic)** ⏩ Relaxes lateral back muscles on the outer curved side, as well as oblique abdominal muscles, and the rib region (hold for at least 10 seconds)
Back relaxation (5)	RELAX +++ FLEX +++ CO-ORD +++	**1.) Look under the belly (Basic)** ⏩ Activates abdominal and lumbar muscles, relaxes back muscles, mobilises small vertebral joints, promotes balance. (hold for at least 10 seconds)

Note: When you are working on relaxation exercises, the dog should be mentally relaxed and you should make sure he is thoroughly warmed up. Stretching is only effective when the muscles are warm!

FLEX ++

RELAX ++

CO-ORD +

Exercise Band 1

1. Plank (Basic)

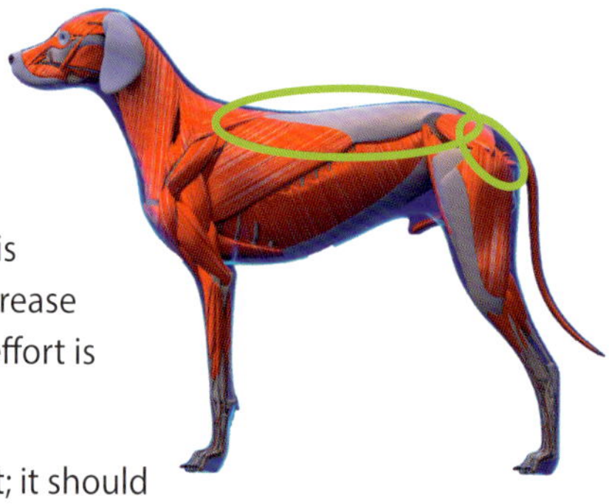

The first step is to teach the dog to stand with his front and back feet on a target. You can then increase the distance between the targets so that more effort is required from the dog.

1. First position a target for the dog's back feet; it should be of sufficient size to allow the dog to be able to stand on it comfortably.
2. Now introduce the dog to the set-up. Lead him over the target, from behind, in a straight line. Click when he touches it with his back feet. At first, it is OK if only one foot touches the target. Feed the dog in this position.
3. Now throw a treat behind the target and click when both back feet are on the target. Feed the dog in this position and repeat several times.
4. Now the front foot target comes into play. Place the target in front of the back foot target, making sure it corresponds to your dog's body length. You want the dog to be able to stand in a relaxed position. Lead him over the targets, again

Gradually increase the distance between the targets.

Equipment TOGU Aero-Step One.

from behind. Do not click until he touches the front target with his front feet; at this stage, one front foot is sufficient. Feed the dog in position.

5. Now click first for both front feet on the target. Keep working until the dog is able to stand on both the front and rear targets, using all four feet.
6. Introduce a verbal cue that means: "stand on the two targets with front and back feet".
7. Reward this intermediate step frequently so that the verbal cue is established.
8. Only now are you ready to start increasing the distance between the targets. Always leave the rear target in place and only change the front target. This makes it easier for the dog to adapt to the changed situation.
9. Proceed in very small steps, i.e. move the front foot target only minimally, at first, and reward the dog at least five times in the new position.
10. Only then are you ready to increase the distance again. If your dog has problems accepting the new distance, take a step back and allow the muscles to take a break.
11. Avoid extremes – the dog should not feel as though he's on the rack – even a small increase in distance can work wonders. Remember, you are also working on relaxing the back muscles. If in doubt, test the effect on yourself.
12. Remember, the dog should hold the position for 10-20 seconds to achieve a relaxing effect.

Tip: If your dog has problems with placing his feet on the targets, use an elevation to clarify the concept of 'go on the elevation and stop'. You can then start working on the exercise. For many dogs, the idea of 'just standing still' is meaningless and they may, therefore, try to offer different behaviours.

FLEX +++

RELAX +++

CO-ORD ++

Exercise Band 1

2. Plank with lowered head (Advanced)

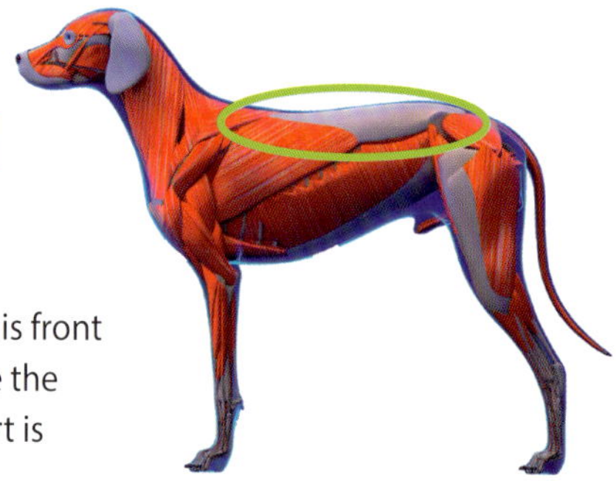

The first step is to teach the dog to stand with his front and back feet on a target. You can then increase the distance between the targets so that more effort is required from the dog.

1. Do not incorporate this more advanced exercise into you training session until the dog is confident performing the basic exercise (see page 134), and fully understands what he needs to do. Prepare yourself. with a clicker, treats and a Post-it.
2. Offer the Post-it as a target, which the dog should touch with his nose. Click and reward. If you have a paw orientated dog, make sure that you hold the target high enough so he is not tempted to work with his paw.
3. Once you have established a nose touch, you can gradually move the Post-it to the floor. Click and reward the steps in between.
4. Now place the Post-it at a slight distance from the dog's front feet; he should be able to reach it comfortably with his nose, without having to move from a standing position.

You can increase the stretching effect by lowering the head. Equipment: TOGU Aero Step One.

5. Consistently click for nose touches on the Post-it. Now you need to focus on reward placement. To encourage the dog to repeat the behaviour, don't feed him at the Post-it, but hold the treat high enough so that he raises his head. This will also prevent him from lying down to eat the treat.

6. After the dog has eaten his treat, he will lower his head to the Post-it. Click the nose touch and, again, feed him in such a way that he has to raise his head.

7. Now work on the duration of the behaviour by delaying the click. This should be done incrementally; do not ask for too much, too soon.

8. If the dog understands what he needs to do, you can introduce a verbal cue, just as he lowers his head to begin the exercise again.

9. Gradually reduce the size of the Post-it. Again, this can be challenging if you progress too quickly. If necessary, feed the dog directly on the Post-it from time to time, to reinforce its value.

10. Now try to omit the Post-it completely. Click your dog at the lowest point of his downwards movement.

11. Remember, the dog should hold the position for 10-20 seconds to achieve a relaxing effect.

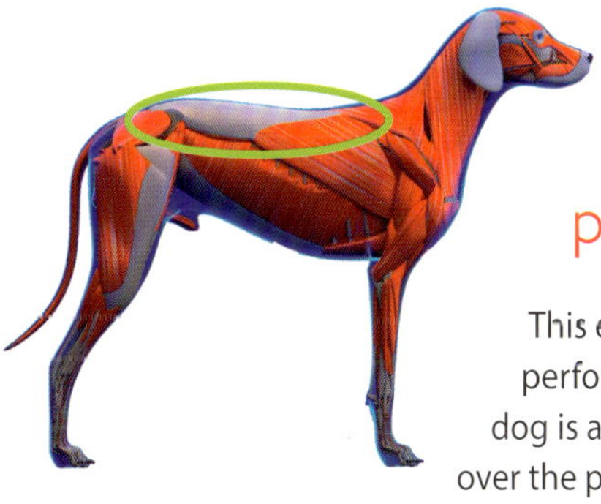

Exercise Band 2

1. Lying belly-length over a peanut ball (Basic)

+++ FLEX

+++ RELAX

This exercise is incredibly relaxing, but it needs to be performed in In a quiet, familiar environment so the dog is able to relax completely when he is stretched over the peanut ball. This exercise is trained with the help of a hand target. Fixating on reward, as a result of luring, or too much trial and error, as in shaping, is counter-productive.

1. You will need a gym ball, or a peanut ball, which is the right size for the dog to stretch over, lengthways, without touching the floor. When choosing treats for this exercise do not use anything that is too stimulating but, rather, something the dog finds interesting, and is able to chew on for longer.

2. Depending on the individual dog, you can work directly with the gymnastic ball/ peanut, or start training with a stool. A calm, confident dog will cope with the wobbly equipment; a more excitable dog will be better off learning the exercise with a stool.

3. Calm dogs: You can use a step-by-step approach to guide the dog over the gymnastic ball/peanut with the help of a hand target. First work on holding the head over the peanut, then on placing one front foot on it, then both. Click and reward each stage, several times, until the dog is proficient. The biggest challenge is placing the back feet. You can secure the gymnastic ball in a stand so that it doesn't wobble; if you are using a peanut, make sure you secure it firmly. Initially, click and reward the dog for lifting his back leg, even if it is only minimal. Work on this gradually until the dog is in position, stretched over the gymnastic ball/peanut, without touching the floor. Bear in mind, you are unlikely to train the full exercise in a single session, so be prepared to take your time. In fact, the dog will gain a better understanding of what he needs to do – and will relax more completely – if you take it slowly. When you eventually achieve the desired position, and the dog is calm and confident, you can gently move the gymnastic ball/peanut back and forth to make it even more relaxing.

4. Excitable dogs: If you have an excitable dog, prepare a set-up with a stool for the dog to climb on, and then secure the gymnastic ball/peanut to it. This prevents the dog from jumping on to, or over, the ball, and risking injury. Click and reward the dog for placing his front feet on the stool, and then progress to moving on to the gymnastic ball/peanut, step-by-step. Click and reward at every stage to bring clarity and calmness to the exercise.

The contact with your body offers the dog the necessary security to hold this position.

Equipment: FitPaws pimpled peanut ball.

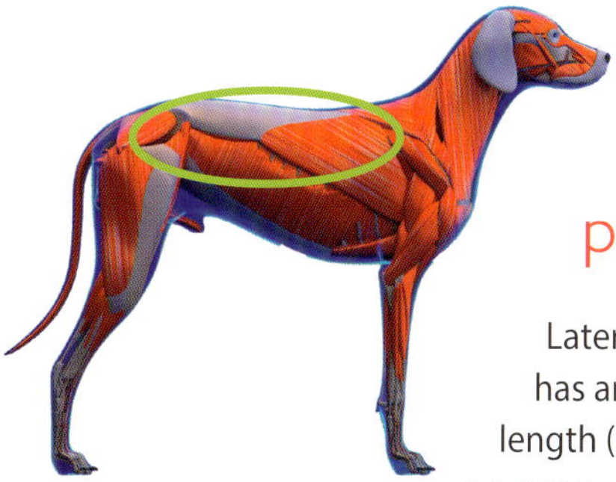

Exercise Band 2

2. Lying sidelong over a peanut ball (Advanced)

+++ FLEX

+++ RELAX

Lateral relaxation, using a gymnastic/peanut ball, has an even greater effect than lying across it, belly-length (see page 137). But on the dog's part, it requires even more trust and confidence. In order to achieve real relaxation, you should first spend time encouraging the dog to lie on his side, without any intervention. For this exercise, you will need some tempting high-value, easy-to-feed treats; a tube of liver paste is ideal.

1. Refresh the hand target by offering it several times and rewarding it directly.
2. Now ask the dog to sit.
3. Decide which side you want your dog to lie on. If you know he has a favourite side, train the other side first. This way, he will later lie on his favourite side as if by magic.

Tip: In this scenario, some dogs struggle with the proximity of the hand target so you may do better with a target stick.

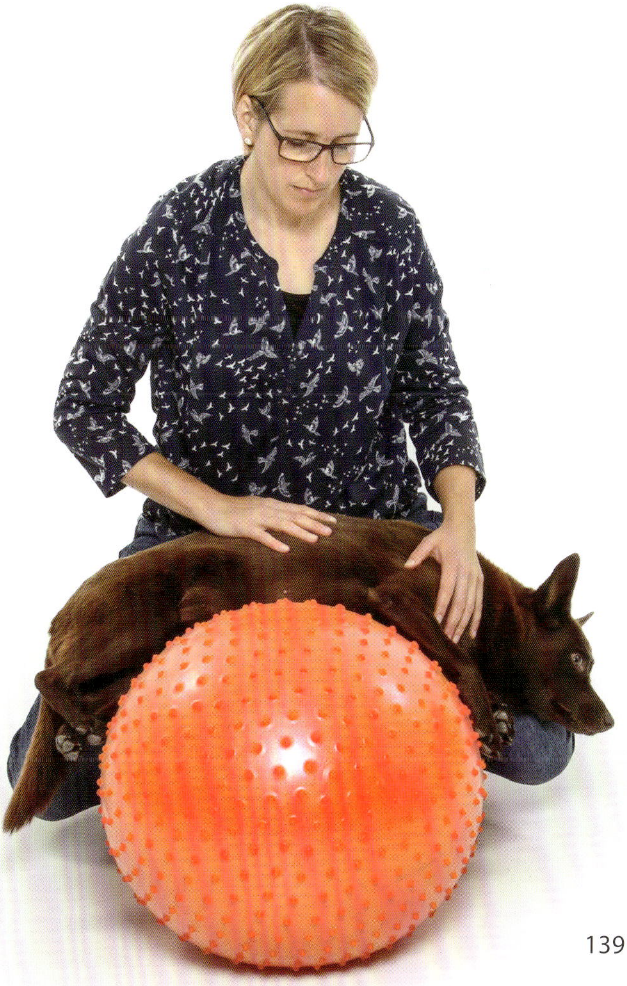

Gently stroke your dog when he is in position, to help him to relax.

Equipment: FitPaws pimpled peanut ball.

4. Offer a hand target to the dog in such a way that he has to turn his head to one side to reach it. Reward this intermediate step and allow the dog lick the liver paste in this position.

5. Proceed gradually so that, at some point, the dog has to bend his upper body to the side to touch the hand target. It is helpful to imagine a semi-circle from tip of nose to tip of shoulder, which your target hand follows. At this stage you will, most likely, need a break.

6. Refresh the steps you have been working on.

7. Now the aim is for the dog to 'tip over' so he is lying flat on his side. This can be achieved by strategic placement of your target hand. However, you will need to experiment to get the desired result.

8. If your dog lies completely on his side, you can introduce a verbal cue. Make sure that you only reward when the dog is lying down quietly, in the exact position you want.

9. Once the dog has linked the verbal cue with the action, you can eliminate your target hand.

10. Now work on duration – the length of time the dog remains lying down –by delaying the click. If necessary, you can support the dog with a chin target ("hold"), a 'keep going' signal, or with some soothing praise.

11. If the dog is happy lying in position, you are ready to advance the exercise. You can now ask him to take up the position on a wobbly surface. Start with something less yielding, such as an old mattress, and then progress to something more wobbly, like a partially filled air mattress.

12. The next big step is imminent: you are now ready to transfer the exercise to the peanut ball. Ask the dog to lie on his side on the peanut. He should now understand the concept of going into this position on a wobbly surface but, if necessary, help him with a hand target. It is a good idea to position the peanut near a sofa, or something of similar height, to prevent the dog sliding down. This is an interim step until the dog understands the exercise and is able to perform it reliably.

13. If the dog is holding his position, and appears completely relaxed, you can move the ball gently back and forth. However, your hands should always remain on the dog to give him the necessary support.

Option: For extremely sporty and confident dogs, you can take the exercise one step further. Ask the dog to sit directly next to the peanut and then, using a hand target, encourage him to tip over into the lying position. To achieve this, the dog must be calm, and the movement must be controlled.

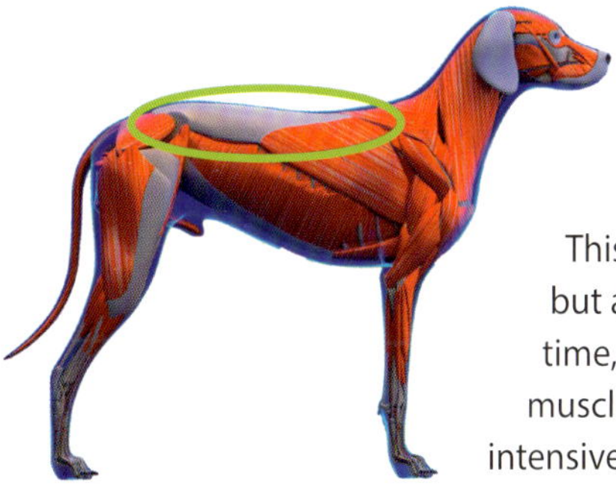

Exercise Band 3

1. Stretching (Basic)

++ FLEX

+++ RELAX

++ CO-ORD

This exercise not only relaxes the back muscles but also stretches the hip extensor. At the same time, depending on the height of the target, the muscles of the forehand are addressed more or less intensively. This exercise can be lured, or taught with a hand target as follows:

1. First position a target ready for the dog. It is intended for his back feet and should be large enough to accommodate both of them comfortably.
2. Now introduce the dog to the set-up. Lead him over the target from behind, in a straight line. Click when he touches the target with his back feet. At this stage, it is OK if only one foot touches the target. Feed in position.
3. Move the dog back over the target, and now withhold the click until both back feet are on the target. Feed in this position and repeat this exercise several times.
4. If your dog is standing securely, with his back feet on the target, you are ready to start the exercise. First, position yourself in front of the dog and offer him a hand target. Reward him for touching it.

Work slowly to ensure the dog's spine is in a straight line.

Equipment TOGU Aero-Step One.

5. Now hold the hand target as an extension of the dog's back line, i.e. so that the spine is in a straight line as he stretches forward. Click and reward for touching the target.

6. Maintaining this imaginary straight line, start to fade the hand target so the dog has to move his front legs to move in your direction. It is important here that the back feet remain on the target, i.e. static. If the dog tries to move his back legs forward you have probably progressed too quickly. If this is the case, simply take a step back.

7. Observe the dog to discover what degree of stretching is possible before his back feet leave the target. Respect this at all costs!

8. Remember, the dog should hold this stretch for 10 to 20 seconds.

9. Over time, the dog may be able to stretch further. However, in this case less is more. Exaggerations are not only inappropriate but also counter-productive as they can lead to tension.

FLEX +++

RELAX +++

CO-ORD +++

Exercise Band 4

1. Knee/flank stretching (Basic)

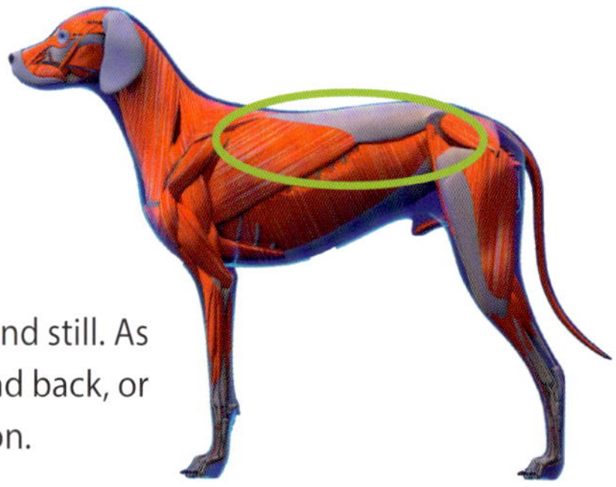

For this exercise, the dog needs to be able to stand still. As an aid, you can place him on a platform, front and back, or alternatively, ask him to stand in the heel position.

1. First of all, reward the dog for standing still.

2. Then ask for a hand target – initially in an arbitrary position.

3. Now, decide which position you want to train first: the stretch to the knee, or the stretch to the flank.

4. Proceed in very small steps, moving your hand target in an imaginary semi-circle from the tip of nose to the knee or flank. Click and reward each intermediate step. This will prevent the dog from getting out of position.

5. The aim of the exercise is for the dog to stretch all the way to his flank or knee, and then stay in that position. You can achieve this by delaying the click as you guide the dog into position.

Tip: Some dogs find it helpful if you place a supporting hand on the thigh, which gives an added sense of security.

6. If the dog is stretching perfectly on one side, train the other. This will, most likely, show which is the dog's preferred side. If there is no medical issue to explain this preference, spend more time training the least favoured side. Then, after a few training sessions, compare his mobility to the better side.
7. Now you can train the stretch to the knee or flank, depending on which one you have not yet trained.
8. When the dog fully understands both stretches, you can combine both options, and work on both sides. Remember, the dog needs to hold each stretch for a minimum of 10 seconds.

The elevations, front and back, make it easier for the dog to stand still.

Equipment: Dog-Stepper.

FLEX +++

RELAX +++

CO-ORD +++

Exercise Band 5

1. Looking under the belly

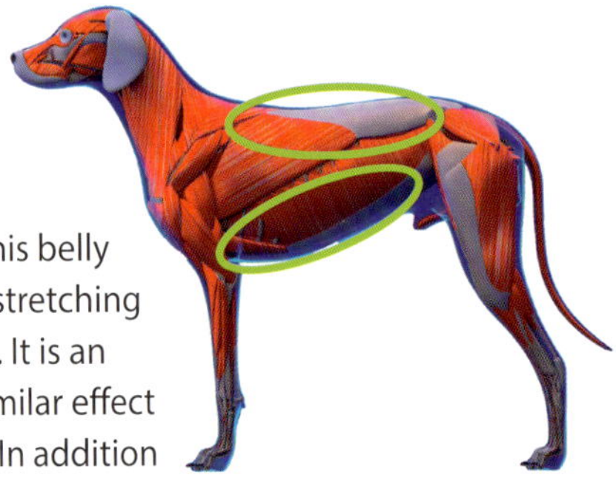

Asking a dog to lower his head and look under his belly is a complex exercise, which requires extensive stretching of the back and neck, and the shoulder muscles. It is an ideal general stretching exercise, which has a similar effect as when you circle your head and then lower it. In addition to the relaxing effect, it is an ideal 'counter' exercise to train the muscles of dogs that do a lot of heelwork – not only to support the position, but also to encourage counter movement.

1. To ensure that the dog remains standing during this exercise, it is a good idea to use an elevation.
2. First of all refresh and reward *plank with lowered head*, (page 136).
3. Now take a Post-it and position it directly underneath the dog by moving your arm between his back legs to indicate the right direction. If the dog is looking to see what you are doing under his belly, click for this initial interest. It is important to feed the dog in this rather unusual position.
4. Now position the Post-it just in front of the dog's back legs. If necessary, you can arouse his interest by pointing at the Post-it or, if necessary, by putting a treat on it. Click and reward the dog for looking towards the Post-it. At this stage it is not anatomically possible for the dog to touch the Post-it. However, you are trying to achieve a rounding of the back as he lowers his head and looks back towards the Post-it. Positioning the dog on an elevation means he cannot sit or to turn round.
5. When you achieve the desired position, you can introduce a verbal cue, and then gradually fade the Post-it by decreasing it in size.

OPTION:
Elevate the front feet only and then ask for the exercise. This will result in a stronger stretch.

The longer the dog holds the position, the more relaxing the effect.

Equipment: Dog-Stepper.

VII. EXERCISES TO ACTIVATE THE BACK MUSCLES

Back muscles: activation		
Category:		**Spine / back muscles**
Training goal:		*Activation of the back muscles, abdominal muscles, lumbar and iliopsoas muscles*
Recommended: *No Go:*		*To maintain and build back muscles* *Sub-acute and acute back problems with changes in the spine* *Consultation with a vet or physiotherapist is highly recommended in case of abnormalities*
Muscular function:		• *Back muscles – Bending, stretching, sideways bending, locks the spine, power transmission from hindquarters to forehand* • *Abdominal muscles – Support movement, antagonist of the back muscles, can tense segmental reflectors (e.g. when back muscles are firm), support breathing, bowel movement, urine flow, labour pains* • *Lumbar, intestinal muscle – Forward swing of the leg in motion, flexor hip (simultaneously turns knee outwards)*

Back activation (1)	CO-ORD ++	*1.) Backing to a target (Basic)* ▸▸ *Training of the short and deep back muscles*
	STRENGTH +++ CO-ORD +++ TACTILE ++	*2.) Climbing stairs backwards (Advanced)* ▸▸ *Addresses mainly short and deep back muscles*
Back activation (2)	STRENGTH +++ CO-ORD ++	*1.) Balancing act (Basic)* ▸▸ *Whole body workout, training of short and deep back muscles, activates abdominal muscles*
	STRENGTH +++ CO-ORD +++	*2.) Balancing act on gyro (Advanced)* ▸▸ *Full body workout, training of short and deep back muscles, activates abdominal muscles, additional extreme coordinative effect*

Back activation (3)	FLEX ++ CO-ORD ++	*1.) Tight spin (Basic)* ▸▸ *Additionally activates oblique abdominal muscles*
	STRENGTH ++ CO-ORD +++ FLEX +++	*2.) Tight spin wobbly/(Advanced)* ▸▸ *Additionally activates oblique abdominal muscles, promotes coordination*
Back activation (4)	STRENGTH ++ CO-ORD ++	*1.) Sphinx nudge (Basic)* ▸▸ *Addresses the entire back and lumbar musculature*
Back activation (5) *Isometric exercises for back activation*	STRENGTH ++ CO-ORD ++	*1.) Isometry: Tripod stand (Basic)*
	STRENGTH ++ CO-ORD + TACTILE +	*2.) Isometry: Pressure from chest towards hindquarters (Basic)*
	STRENGTH +++ CO-ORD ++ TACTILE +	*3.) Isometry: Diagonal pressure over shoulder towards hindquarters (Basic)*
	CO-ORD +++	*4.) Isometric exercises/wobbly (Advanced)*
Back activation (6)	CO-ORD +++ STRENGTH ++ RELAX +	*1.) Mountain goat for the back (Basic)* ▸▸ *Whole body workout: training of short and deep back muscles, activates also the abdominal muscles*

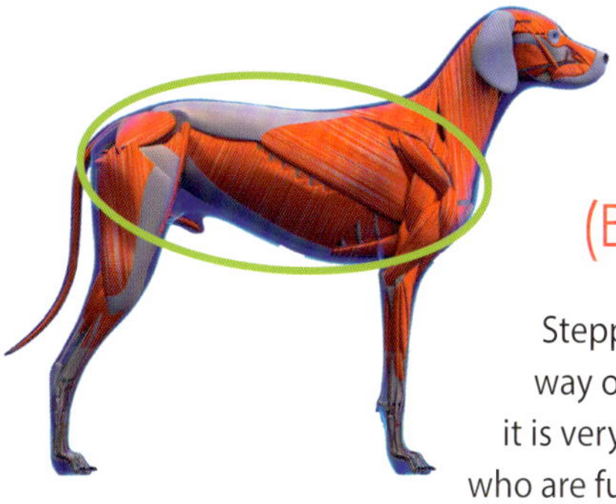

Exercise Band 1

++ CO-ORD

1. Backing to a target (Basic)

Stepping back on to a low target, or a step, is a great way of improving your dog's co-ordination. However, it is very strenuous and is only recommended for dogs who are fully fit. Although the back legs are doing the work, the front legs are bearing the weight and thus are put under a considerable amount of strain.

1. First of all, ask for, and reward, the dog for walking back.
2. Now start by placing a small elevation behind the dog; it should be of sufficient size to accommodate his back feet comfortably. You can use a dog-stepper element, or the lid of a shoebox, for example. Make sure that the elevation is secure and cannot slip.
3. Ask the dog to walk backwards again. Initially, he may be puzzled by the 'obstacle' so it is important to click and reward the moment he touches it.
4. Repeat the walk back several times, but delay the clicks until the dog is actually standing on the elevation with at least one back foot.
5. Keeping working in this way until he is standing on the elevation with both back feet.

For this exercise to be successful, the dog needs to learn to stand completely still.

Equipment: TOGU Senso balance cushion.

CO-ORD +++

STRENGTH +++

TACTILE ++

Exercise Band 1

2. Climbing stairs backwards (Advanced)

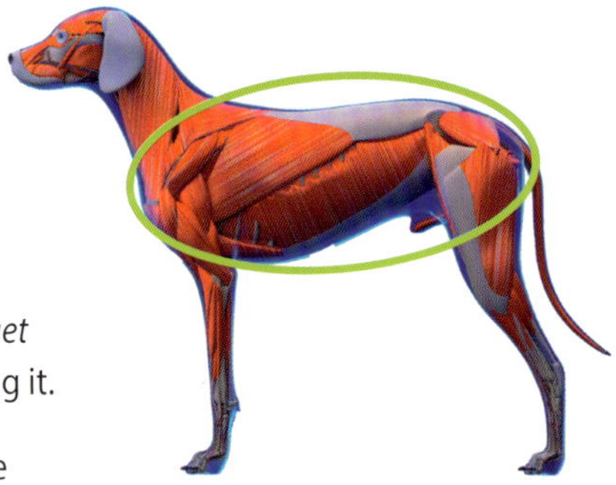

1. Refresh the basic exercise, *sbacking to a target* (see page 147), and reward the dog for doing it.
2. Now you can add a second 'step'; this can be two dog-stepper elements, or a shoebox. Encourage the dog to climb backwards from the first elevation to the second by using your cue to reverse.
3. If the dog can do this reliably, introduce a verbal cue that allows him to adjust to stepping up backwards.
4. Generalise the exercise in another place. Use 'natural' obstacles, such as real stairs or stacks of wood. You can ask for this exercise on walks.

Initially, position the dog close to the stairs so he only has to walk back a short distance before he finds the first step.

Equipment: Stair-like construction.

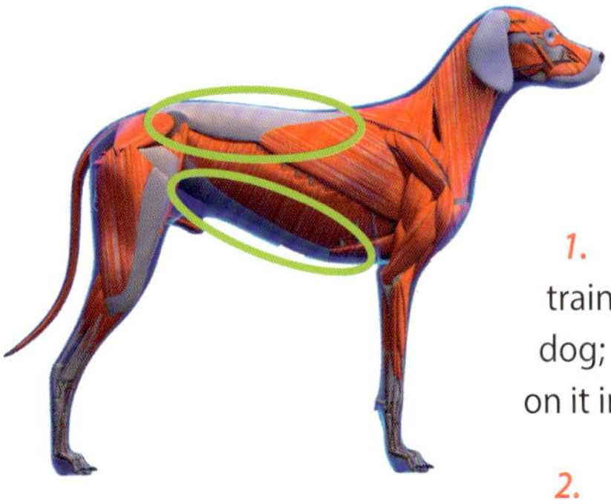

Exercise Band 2

++ CO-ORD

+++ STRENGTH

1. Balancing act (Basic)

1. First, set up a large target, or platform, in your training area. The size depends on the size of your dog; it should be large enough for the dog to stand on it in a relaxed manner.
2. Now introduce the dog to the set-up. Click and reward him for showing initial interest. Next, lead him across the platform, from behind, in a straight line. Click when he stands on it and feed him in position.
3. Work on his understanding of the equipment by luring him so that he removes one foot while he is getting his treat, and then must 'sort himself out' and return, independently, to the correct position for the next click.
4. Now throw the treat behind the platform so that the dog has to get off the equipment to retrieve it. He can now approach the platform from behind and get back on to it. Work on this gradually, making it more difficult, and throwing the treat sideways.
5. Make sure you click and reward the dog for standing, not when he is moving. This will help him to understand that he must be still before he is rewarded.
6. If he is happy to do this exercise on a stable elevation, you can now try it on a wobbly surface. Start with an object that moves only minimally and then move on to increasingly wobbly surfaces.

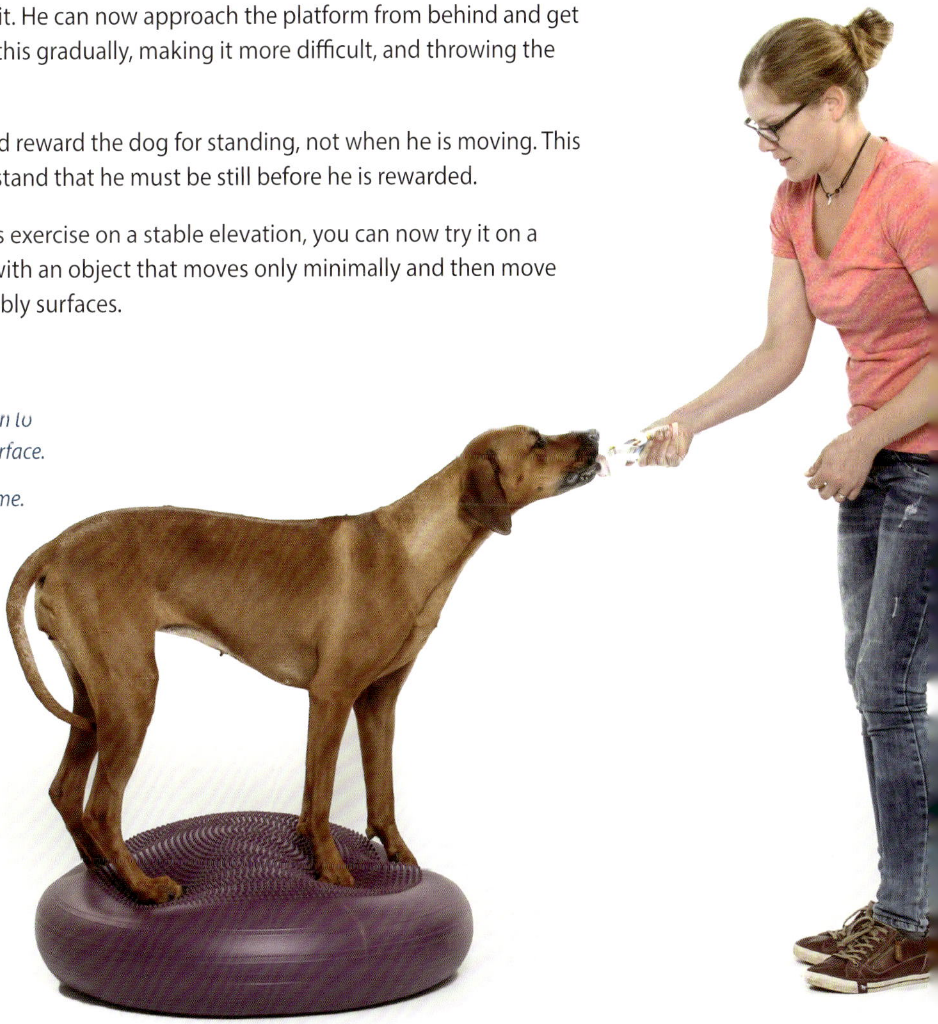

With practice, your dog will learn to balance himself on a wobbly surface.

Equipment: TOGU Dynair Extreme.

CO-ORD +++

STRENGTH +++

Exercise Band 2

2. Balancing act on a gyroscope (Advanced)

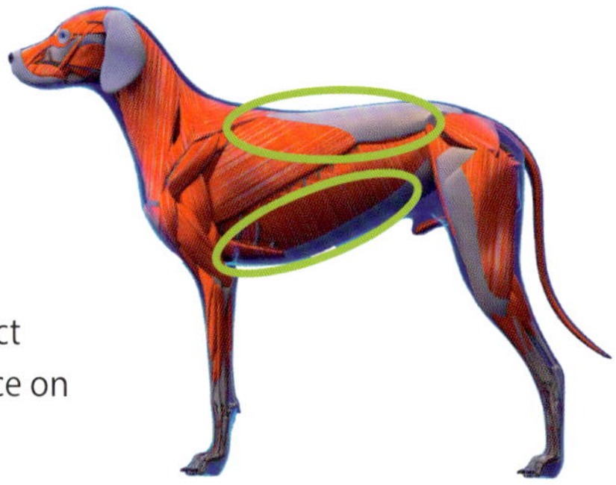

As an advanced version of the basic balancing act (see page 149) , you can train your dog to balance on a gyroscope/spinning top.

1. You will need to secure the gyroscope before you start training. You can do this by placing cushions on both sides so that the top only moves in two directions and wobbles less.
2. Lead the dog on to the equipment, from behind, using a hand target. Reward him for placing his front feet on the gyroscope.
3. If the dog is comfortable with his front feet on the equipment, guide him forward so that he places his back feet on the gyroscope as well. Click and reward.
4. When the dog is standing with all four feet on the gyroscope, you can make it wobble a little more.
5. If the dog still feels comfortable, you can gradually remove the cushions from underneath. Reward the dog for each intermediate step.

Here, you can see how well the dog is balancing himself.

Equipment: Therapy spinning top.

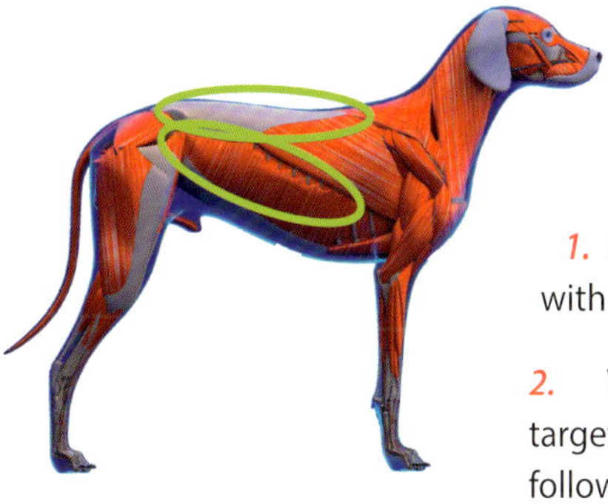

Exercise Band 3

1. Tight spin (Basic)

++ FLEX

++ CO-ORD

1. Refresh your hand target or, if you prefer, work with a target stick.
2. Working step-by-step, use your hand target, or target stick to describe a circle, which the dog will follow. Initially, mark and reward even the slightest bend to help the dog understand what is required.
3. Keep practising – and rewarding – until the dog can complete a 360 degree turn.
4. Try to keep the rotation as tight as possible to ensure that the muscles are responding appropriately. Make sure the dog is working calmly; fidgeting is counter-productive. If this happens, take a step back and reward the dog after 90 degrees, or after 180 degrees, for example.
5. If the dog is achieving a tight spin, you can introduce a verbal cue and gradually fade the hand target or target stick.
6. Now work on a tight spin on the other side, following steps 1-5.
7. Pay attention to how the dog executes his tight spins. Which side is easier for him? Is he trying to rush more on one side than the other? The aim is to make him equally flexible on both sides.

This dog is learning to turn tightly with the help of a cone.

FLEX ++

CO-ORD +++

STRENGTH +++

Exercise Band 3

2. Tight spin/wobbly (Advanced)

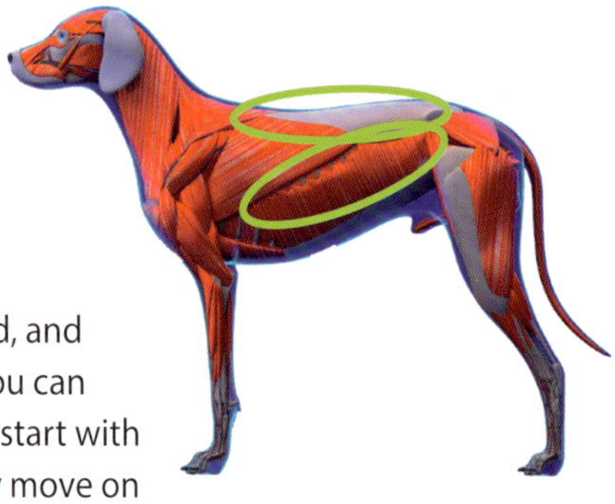

Once your dog can do a tight spin on the ground, and performs it cleanly to the left and to the right, you can transfer the trick to a wobbly surface. To do this, start with an old mattress, for example, and then gradually move on to increasingly wobbly surfaces, such as a peanut ball.

Here you can see the dog turning on a peanut ball.

Equipment: FitPaws pimpled peanut ball.

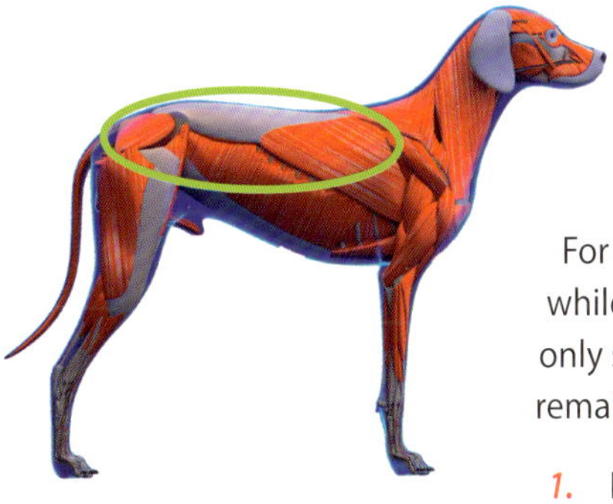

Exercise Band 4

1. Sphinx nudge (Basic)

++ CO-ORD

+++ STRENGTH

For this exercise the dog learns to nudge a target while he maintains a sphinx position. He should only stretch forward using his back; his feet should remain in place.

1. Refresh your hand target or, if you prefer, work with a target stick.
2. Ask the dog to lie down, making sure he is in the correct sphinx position.
3. Now offer the target so close to his nose that he does not need to move his body forward, he can reach it by stretching his head forward.
4. Gradually offer the hand target further and further away so the dog has to stretch to the maximum. If he attempts to crawl, or stands up, you are asking for too much and will need to take a step back.
5. This exercise works through tension and relaxation, which means that you should reward the dog when he is in the relaxed starting position. Further nudging of the target creates a kind of rocking movement, which is particularly visible on the croup.

Make sure the dog does not attempt to crawl when you are working on this exercise.

Exercise Band 5

Isometric exercises

Isometric exercises represent strength training, without changing the length of the muscles, which means they are also safe to use for rehabilitation. However, you should always check with a health professional before starting on an exercise programme if your dog is recovering from injury, or from surgery.

When a dog is engaged in isometric exercises, he tends to behave more passively or is reacting to pressure touches. It is, therefore, advisable to carry out these exercises in a different setting to the equipment training to make it easier for your dog to understand the difference.

CO-ORD ++

STRENGTH ++

Exercise Band 5

1. Isometry: Tripod stand (Basic)

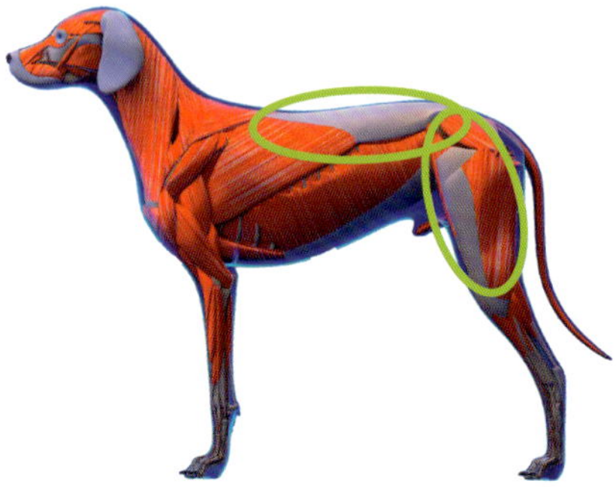

1. Ask the dog for his paw.
2. Make sure you keep his paw close to his body. This is very different from *lifting front legs* (see page 126)
3. Gradually extend the duration of the paw lift by delaying the click.
4. The aim of this exercise is for the dog to be relaxed and hold the position for at least 10 seconds.

The dog's back must compensate for the weight shift which results from giving a paw.

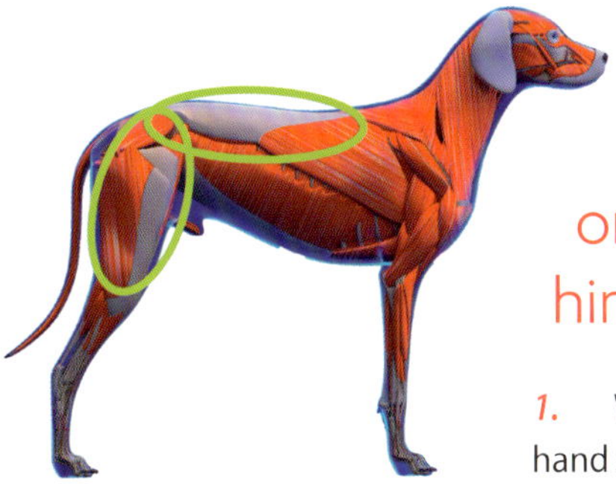

Exercise Band 5

2. Isometry: Pressure on the chest towards the hindquarters (Basic)

+ CO-ORD

++ STRENGTH

+ TACTILE

1. With the dog in a stand, place the flat of your hand across his chest. Click and reward him for accepting this without changing position. Initially, you may need to take it more slowly and reward him for standing still as your hand approaches, and then progress gradually, until your hand is on his chest.

2. If the dog accepts your hand touch, you can start to build up minimal pressure for very short periods. Click into the pressure, i.e. when the dog builds up counter-pressure. If he backs off, be more gentle, and proceed more slowly.

3. Apply the pressure for at least 10 seconds, and then, gently, release it. If you continue for too long, the dog will, literally, fall on you.

4. To facilitate learning, it is helpful to introduce a specific verbal cue so the dog understands the exercise does not require active movement, but is to do with a passive counter measure. As a rule, one cue – e.g. "iso" – is sufficient for all isometric exercises, unless you want to name individual body parts separately in order to use them as targets later on in training.

Tip: If your dog finds it difficult to build up counter-pressure, or to sustain it, it may help if you start the exercise working with a chin target.

The dog must create a counter-pressure with his hindquarters and back.

CO-ORD ++

STRENGTH +++

TACTILE +

Exercise Band 5

3. Isometry: Diagonal pressure over the shoulder towards the hindquarters (Basic)

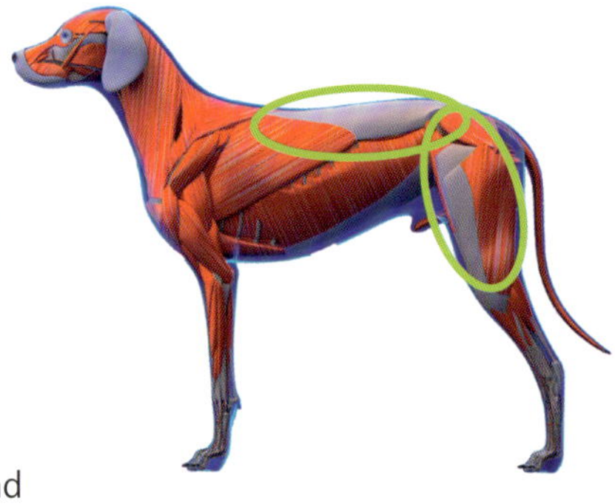

1. Similar to the previous exercise, first click and reward the dog for accepting your hand on his shoulder. He should be relaxed and remain standing still.
2. Now slowly build up pressure and reward the dog's counter-pressure.
3. The dog needs to respond with counter-pressure for at least 10 seconds. Here, the long back extensor is addressed intensively via the diagonal.
4. Now change sides and train from the other shoulder. As a variation, you can also do this isometric exercise on the hip. In this case, you train the thigh muscles in addition to the back muscles, whereas in the described version you are working on the back muscles and, in addition, strengthening the shoulder area.

Think of the pressure being applied diagonally to the opposite hip.

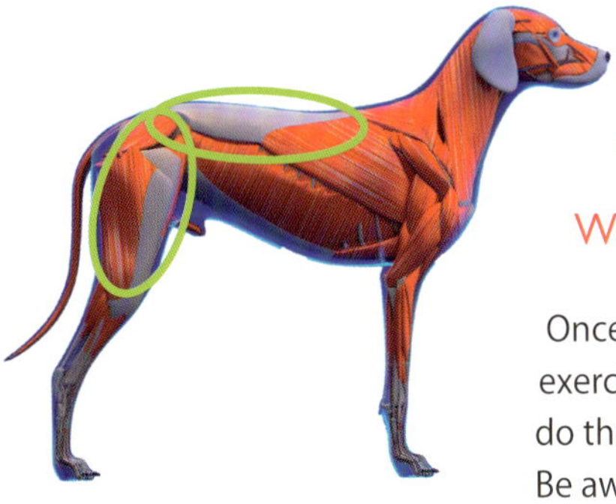

Exercise Band 5

+++ CO-ORD

4. Isometric exercises/ wobbly (Advanced)

Once your dog has understood the isometric exercises, you can intensify them by asking him to do them on a wobbly surface. Proceed with caution. Be aware that you have now changed one of the criteria in training and it is therefore sensible to go back a step so you can reward as he learns. Build duration gradually.

The dog should learn to balance on wobbly equipment before being asked to lift a leg.

Equipment: FitPaws FitBone.

Exercise Band 6

1. Mountain goat for the back (Basic)

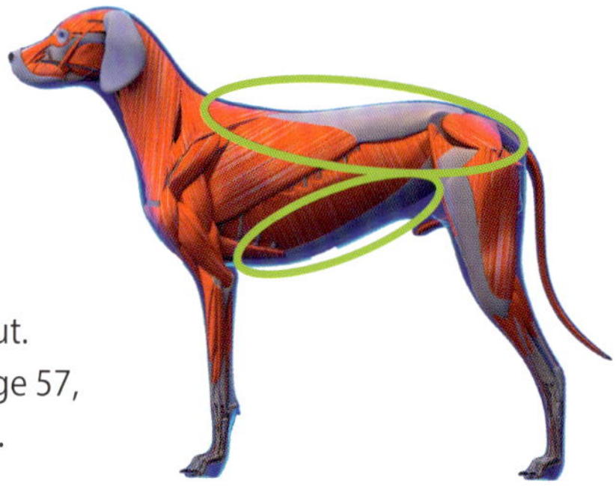

The *mountain goat* exercise is a full-body workout. Train it according to the exercise outlined on page 57, and pay particular attention to a calm execution.

The mountain goat exercise is brilliant for activating the back.

Equipment: Ikea stool (large).

VIII. EXERCISES TO STRENGTHEN THE PELVIC FLOOR

Pelvic floor		
Category:		***Pelvic floor***
Training goal:		*Strengthening of the pelvic floor*
Recommended:		*For dogs with weak muscles of the pelvis. This can be helpful for incontinence issues if other health causes have been ruled out in advance (e.g. cystitis, disc problems)*
Muscular function:		*Secures the position of the abdominal and pelvic organs, supports the urethra and the sphincter muscle of the bladder and helps to close the anus*
Pelvic floor ①	STRENGTH ++ FLEX ++	*1.) Lateral sit-ups (Basic)* *Side sit-ups lying on the back. Stabilisation through the handler's legs* ⏩ *Strengthens the pelvic floor muscles and activates the abdominal muscles*
Pelvic floor ②	STRENGTH + CO-ORD ++	*2.) Side stretch/wobbly (Basic)* *Side stretches sitting on a wobbly surface* ⏩ *Strengthens the pelvic floor muscles, coordinative effect*
Pelvic floor ③	STRENGTH ++	*1.) Bounce (Basic)* *Bounce from a sit to a hand target* ⏩ *Strengthens the pelvic floor muscles*
	STRENGTH +++ CO-ORD ++	*2.) Bounce/ wobbly (Advanced)* *Bounce from a sit to a hand target – sitting on a peanut ball* ⏩ *Strengthens the pelvic floor muscles, co-ordinative effect*
Pelvic floor ④	STRENGTH +	*1.) Squats (Basic)* *Keep the transition between sitting and standing short* ⏩ *Strengthens pelvic floor muscles, very strenuous*

Pelvic floor (4)	STRENGTH + CO-ORD +	***2.) Squats/wobbly (Advanced)*** *Transition between sitting and standing on a wobbly surface* *▸▸ Strengthens the pelvic floor muscles, co-ordinative effect*
Pelvic floor (5)	STRENGTH + CO-ORD +	***1.) On the stepper (Basic)*** *Stepping back on to a stool, alternating back feet* *Strengthens the pelvic floor muscles, co-ordinative effect on hind limbs*

Dogs, like us, have pelvic floor muscles and, just like us, strengthening these muscles can have a positive effect on incipient incontinence.

To get a feel for the tension and relaxation of your dog's pelvic floor, proceed as follows:

When the dog is standing, feel the ischial tuberosities. These are the bony structures that are located on the right, and left, under their tail. From there, move one finger on either side towards the base of the tail. There, you will feel a small hole, or depression, which is formed by the pelvic floor muscles.

If possible, recruit an assistant when you are working on *bounce – basic* (page 163) and *squats – basic* (page 165) so that you leave your fingers in position during the exercise and can feel the tension in the muscles. Obviously you must ensure this is not uncomfortable for your dog.

FLEX ++

STRENGTH ++

Exercise Band 1

1. Lateral sit-ups (Basic)

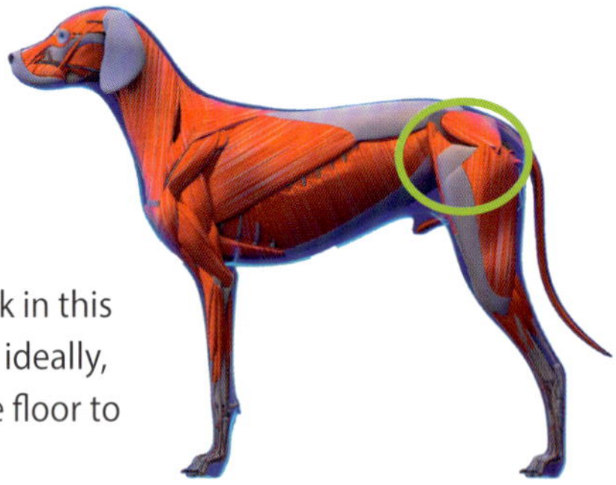

For this exercise, the dog must feel relaxed and comfortable so that he is willing to lie on his back in this vulnerable position. Choose a familiar location – ideally, your home –and, if necessary, place a mat on the floor to ensure the surface has minimal impact.

1. Make yourself comfortable on the floor and spread your legs in a V shape. Ask the dog to join you. For option 1, you will need a Post-it as a target.

2. First you ask for a down and reward this generously. If your dog knows how to lie on his back, you can ask for this. Otherwise, pet your dog until he can be manoeuvred on to his back in a relaxed manner.

Option 1: nose/hand target:

1. As an introduction use a Post-it and mark and reward the dog, a few times, for touching it. You can also use a hand target. In both cases, offer the target in such a way that the dog has to stretch minimally in order to reach it.

2. Slowly but steadily, move the Post-it towards the dog's knee. Proceed in very small steps and don't be in a hurry. After touching the Post-it a total of ten times, take a break. This training is both physically and mentally tiring for your dog.

3. When you recommence training, start again from the beginning. Imagine a person doing sit-ups – this is what you are aiming for when the dog stretches to touch his knee with his nose. Focus on one side only.

4. When the dog is proficient on one side, you can train the other side. Reward him for working calmly; erratic movement is counter-productive.

5. Now you can ask the dog to do sit-ups on alternate sides.

Option 2: Food lure:

Proceed in the same way as for option 1 but, this time, you have food, which you should hold in a closed fist. When the dog follows your fist as you move it towards his knee, you can let the food fall out. This means the dog focuses on what he needs to do rather than mugging you for the treats. Assess his progress and, if necessary, swap to higher, or lower, value food depending on whether the dog needs to be motivated, or whether he is too impetuous.

Work slowly as the dog needs to build confidence in order to perform this exercise.

CO-ORD ++

STRENGTH +

Exercise Band 2

1. Side stretch/wobbly (Basic)

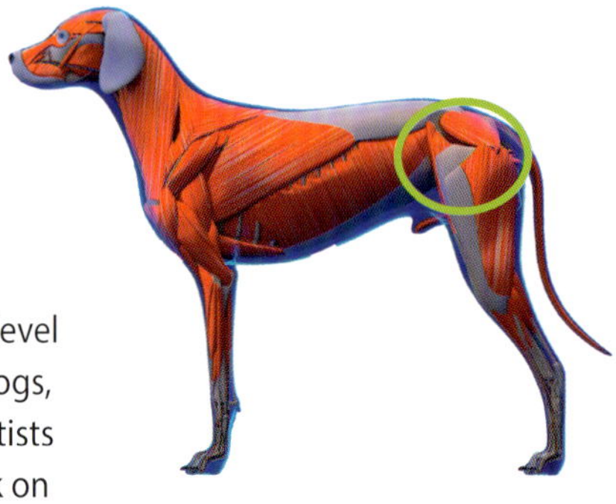

You can adapt this exercise to the performance level of your dog by varying the surface. For novice dogs, start on a large cushion; experienced balance artists can use a peanut ball. However, before you work on a wobbly surface, you need to train it on the floor so the dog understands what is required.

1. Ask the dog sit in front of you.
2. Now offer him a hand target, making sure that he remains seated when touching it. To achieve this, hold your hand directly over his head so that he does not need to thrust forward. Click and reward the hand touch.

Here you can see the stretch to the left.

Equipment FitPaws: Pimpled peanut ball.

3. Now move your hand on an imaginary semi-circle, one side at a time. Click and reward each intermediate step as you gradually move your hand.
4. Try to move your hand in such a way that the dog not only has to stretch the neck, he also has to stretch his front paws, as though he is standing on tiptoe. You also need to make sure he is sitting correctly at the back, i.e. his hind legs do not turn outwards, but lie close to his body. If he turns his hind legs outwards, interrupt the exercise briefly and re-set his position.
5. When the dog is working well on one side, you can move on to training the other side.
6. When he is proficient on both sides, you can move the exercise to a wobbly surface. This is when you will bring the pelvic floor muscles into play. Make sure you always train both sides.
7. Now extend the duration of the exercise.

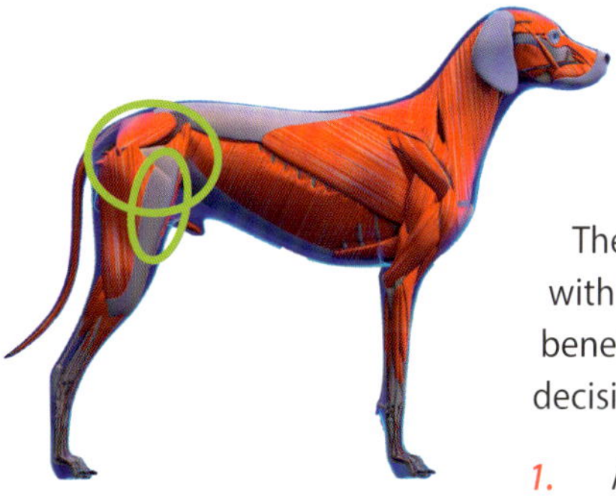

Exercise Band 3

++ STRENGTH

1. Bounce (Basic)

The first training steps of this exercise correspond with those of training 'beg'. However, in order to benefit the pelvic floor, it is the bounce that is the decisive element.

1. Ask the dog to go into a down.
2. Now offer him a hand target, just above the tip of his nose, so that he has to stretch upwards up to reach it. Click and reward.
3. Repeat this several times.
4. Now, gradually, start to hold the hand target higher so the dog learns to push off with the forehand to reach it. Make sure he doesn't stand up; his bottom must stay on the floor and he should sit parallel, i.e. his hind legs must not turn outwards. Observe closely and only click and reward when you can see him pushing up from his rear, keeping his legs in parallel position. If he stands up, or slides out with his hind legs, interrupt the trick briefly and ask him to sit up straight again. Also, make sure he doesn't try to use you as a support. You can prevent this by standing an appropriate distance from the dog.
5. Proceed with caution. This exercise is very strenuous and the dog must first find his balance. Give him time to do this.

The dog should not go into a 'beg'; a slight bounce is all you are looking for.

CO-ORD ++

STRENGTH +++

Exercise Band 3

2. Bounce/wobbly (Advanced)

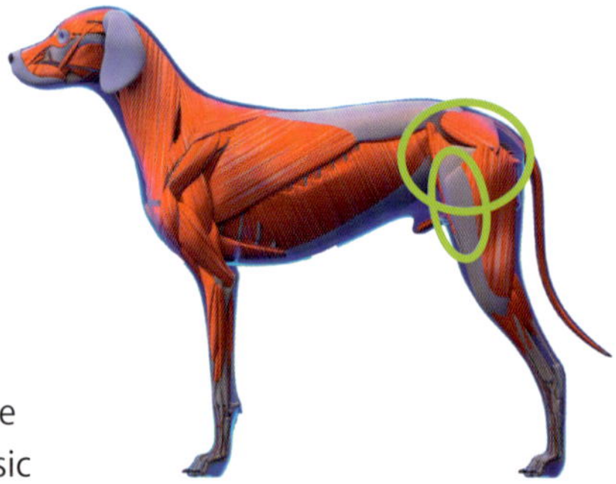

If you have trained your dog to bounce on level ground (see page 163), you can now progress the exercise and train it on a wobbly surface. The basic exercise benefits the pelvic muscles; the advanced version intensifies the effect.

You can choose whether you want to position the dog's hindquarters on an air cushion, or whether he should stand, in entirety, on a peanut ball. The peanut represents the greater training challenge. Its rounded shape means the dog cannot thrust upwards from the resting part of his leg and, therefore, has to use his pelvic muscles more intensively.

To do this exercise, your dog needs to be an expert in finding his balance.

Equipment: FitPaws Donut with ring.

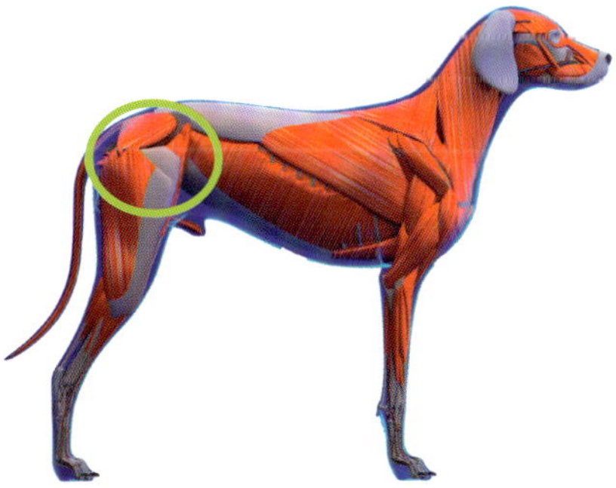

Exercise Band 4

+ STRENGTH

1. Squats (Basic)

The change of position from sitting to standing is well known to sports dogs. It is the transition between the two positions, i.e. neither sitting nor standing, but the movement in-between, that has a significant effect on the pelvic floor.

If you want your dog to perform a perfect position change, outside the realm of fitness training, it is a good idea to create a special fitness training atmosphere so he

TIP: If your dog tends to step forward, practise on an elevation. In this way you can position him in a sit and, when you ask him to stand, he has to remain static otherwise he will have to step off the elevation.

can differentiate between the two exercises. You can do this by changing location, and by using different equipment and verbal cues.

Option 1: Static:

1. Position the dog in front of you in a sit. Offer a hand target at a distance where he has to stand up to reach it. Make sure that he only has to stand up and does not need to step forward. Click the movement – the transition between sitting and standing – and not the end product, i.e. the stand. Ideally, you would feed the dog as he is moving, but this can be a challenge in the early stages of training. If this is the case, feed him when he is sitting or standing, and use this as your starting position to repeat the exercise.

2. Repeat, marking and rewarding this intermediate movement several times.

When your dog has perfected the basic exercise on level ground, you can progress to the advanced version (see page 164) and train it on a wobbly surface.

Equipment: TOGU Ballanza Ballstep

3. Now try to achieve a pause in the intermediate movement. You can only achieve this pause if the dog understands that he will be rewarded for the intermediate movement. If this is the case, delay the click for a millisecond. If he holds position, without standing up or sitting down, reward him generously.
4. To build duration, keep on delaying the click.
5. Do not overdo it – this position is very tiring!

Option 2: Via walking backwards

1. Position the dog in a stand in front of you first. Ask him to walk backwards. Reward him for this.
2. Next, position the dog in a down in front of you. Now it all comes down to good timing. Ask the dog to walk backwards and click the exact moment he lifts his bottom from the floor and straightens his hind legs. Reward the first moment of getting up, i.e. the start of walking backwards.
3. Now ask for a sit, and click the movement of getting up, not the end product.
4. Repeat this transition several times.
5. Now work towards a pause in the movement by delaying the click.
6. Watch out for signs of fatigue!

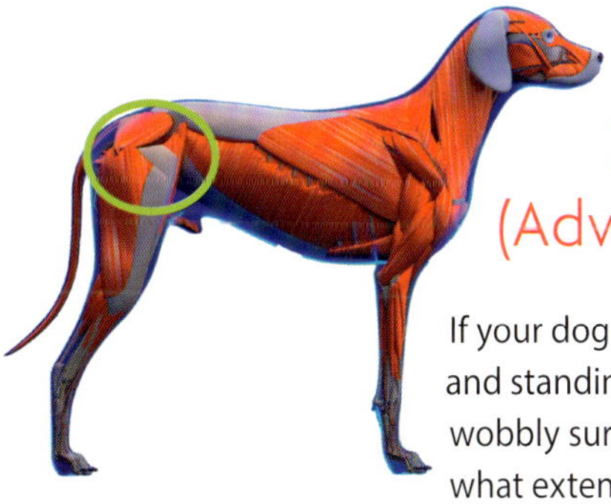

Exercise Band 4

2. Squats/wobbly (Advanced)

+ CO-ORD

I STRENGTH

If your dog can safely make the transition between sitting and standing on level ground, you can transfer this to a wobbly surface. However, proceed carefully and test to what extent your dog can do this safely and comfortably.

CO-ORD +

STRENGTH +

Exercise Band 5

1. On the stepper (Basic)

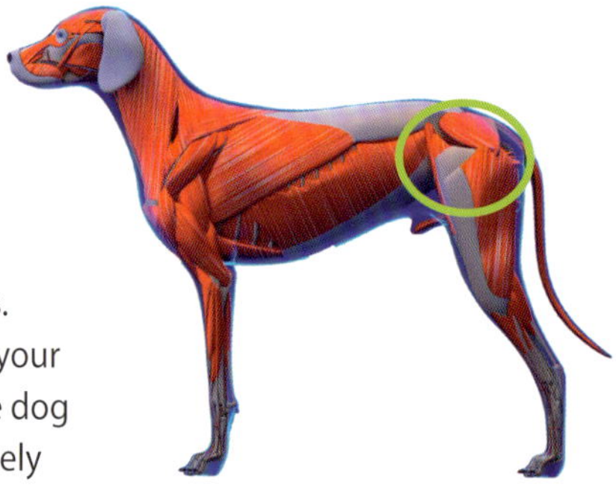

Option 1: Stairs

1. You will need two stools of different heights. The smaller stool should be no higher than your dog's hock, the larger stool should allow the dog to stand comfortably and not to be excessively elevated. A 45-degree angle should be the maximum.
2. Place the stools, one behind the other, in such a way that the dog perceives them as stairs that need to be climbed. Make sure that the distance between the stools is correct, so that the dog finds it easy to stand on both of them.
3. Lead the dog to the stools. The easiest way to do this is to walk backwards encouraging the dog to follow you. If necessary, work with visual boundaries so that he can orientate himself.

The aim is for the dog to place one back foot on to the small stool and then return it the floor. He should then change sides.

Equipment: Ikea stool (large and small).

4. Reward the dog's first attempts to climb the stools. Remember not to expect the end result immediately.

5. If he is happy to step on to the stools, you can progress the exercise.

6. Move the smaller stool slightly to the side. Now it gets interesting, and you will need to observe the dog very closely. The moment he places one back foot on the lower stool, mark, and reward generously.

7. The greatest effect on the pelvic floor is achieved when the dog raises and lowers his back leg. You can achieve this by using a reverse cue, provided the dog knows how to walk backwards. If your dog does not know this, just wait a little. The position is very strenuous and, sooner or later, he will put his foot back on the floor, which you can then reward.

Option 2: Small stool first

1. With this option, the dog learns to stand on both stools rather than focusing on a single back leg, which was required for option 1.

2. Start with the lower stool, and encourage the dog to walk back towards it. As soon as he touches the stool with his back feet click and reward.

3. First, consolidate standing with both back feet on the low stool because this is the critical point for the next step of the exercise. If the dog fully understands what is required, you can proceed with the training.

4. Next add the higher stool. Place it in front of the low stool at the appropriate distance. Now ask the dog to sit over the stools. He will want to position himself with his back end on the low stool and with his feet on the higher stool. Click and reward.

5. Reward this intermediate step several times.

6. Now move the lower stool out of line so that when the dog approaches, in the same way as before, he will only be able to accommodate three legs – two front feet on the higher stool, and one back foot on the lower stool. Click and reward this generously. This step is easier to teach if you can recruit an assistant to help you.

IX. EXERCISES TO TRAIN BODY AWARENESS AND CO-ORDINATION OF THE HIND LIMBS

Body awareness/co-ordination of hindquarters and training of hind limbs		
Category:		**Hind limbs**
Training goal:		*Improvement of body awareness and co-ordination of hindquarters*
Recommended:		*For dogs with poor rear end awareness*
Muscular function:		*The gluteal muscles (croup muscles), anterior and posterior thigh muscles, the M.gastrocnemius (calf muscles) and, when walking sideways, the M.Biceps femoris (biceps femoris) are primarily trained. Depending on the way the dog walks backwards, the back, abdominal and M-iliopsoas (ilio-lumbar muscles) are also trained*
Co-ordination of the hind limbs ①	CO-ORD ++ TACTILE ++ FLEX +	***1.) Backing down an alley (Basic)*** *Trains all the above mentioned muscles*
	CO-ORD +++ TACTILE ++ FLEX +	***2.) Backing on to a sloping board (Advanced)*** *Trains all the above mentioned muscles; additional co-ordinative effect from using the sloping board*
	CO-ORD +++ TACTILE ++ FLEX +	***3.) Backing through an 'L' to stairs (Professional)*** *Trains all the above mentioned muscles, also major co-ordinative effect*

Co-ordination of the hind limbs ①	CO-ORD +++ FLEX ++ STRENGTH +	*4.) Backing down an alley, plus side-stepping poles (Professional)* ▶▶ *Side-stepping additionally strengthens the training of the lateral thigh muscles*
Co-ordination of the hind limbs ②	CO-ORD ++ STRENGTH + FLEX +	*1.) Backwards-forwards over a pole (Basic)* ▶▶ *Good co-ordination and mobility training of the hind limbs. Especially useful for dogs who do not lift their feet high enough, or prefer to walk in pace instead of trot*
Co-ordination of the hind limbs ③	CO-ORD ++ TACTILE + FLEX +	*1.) Finding a back foot platform and holding position (Basic)* ▶▶ *Improves awareness of the hindquarters (also single limb)*
	CO-ORD +++ STRENGTH ++ TACTILE +	*2.) Sit-stand on wobbly back foot target (Advanced)* ▶▶ *Promotes co-ordination and back end awareness*
	CO-ORD +++ TACTILE ++ FLEX +	*3.) Back feet on paw pods (Basic) and front foot change (Professional)* ▶▶ *Promotes co-ordination and back end awareness*
Co-ordination of the hind limbs ④	CO-ORD ++ TACTILE ++ FLEX +	*1.) Placing a back foot on a target (Basic)*

Co-ordination of the hind limbs (4)	CO-ORD ++ STRENGTH +++ TACTILE +	***2.) Lifting alternate back feet (Advanced)***
	CO-ORD +++ STRENGTH ++ FLEX +	***3.) Lifting alternate back feet/wobbly (Professional)*** ▸▸ *Promotes or improves body awareness, and also balance*
Co-ordination of the hind limbs (5)	CO-ORD +++ FLEX ++ STRENGTH +	***1.) Reverse slalom (Basic)*** ▸▸ *Trains co-ordination and also the back muscles*

FLEX +

CO-ORD ++

TACTILE ++

Exercise Band 1

1. Backing down an alley (Basic)

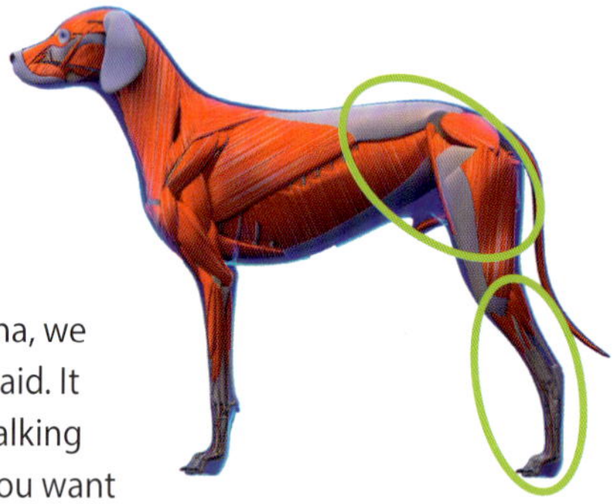

Similar to groundwork performed in a horse arena, we work with an alley, which offers the dog a visual aid. It also separates the exercise from independent walking backwards, as in freestyle, for example. Ideally, you want to work with two different types of treats: high value and low value.

1. With the help of a dog-stepper, or a cavaletti set, build an alley slightly wider than the width of your dog. You will also need a mat and a chair. The chair should be positioned in the middle of the alley.
2. Place the mat in front of the chair so that the distance is equivalent to the length of your dog's body. The dog should automatically step on to the mat, with his back feet, when he is positioned directly in front of you.
3. Now lead the dog, from behind, through the alley and sit down on the chair. Readjust the mat if necessary.
4. Place a low-value treat under the chair in such a way that your dog has to step off the mat with both back feet to get the treat. Now it gets exciting:

5. With a clicker and high-quality treats at the ready, mark the moment the dog moves backwards from his position and 'must', thereby, step back on to the mat.
6. To cement the behaviour of stepping back on to the mat, repeat, at least 10 times.
7. Now move the chair to alter the distance between the chair and the mat. Important: always move the chair. The mat remains in the same place, as the dog is using it as a target. Proceed in stages so that the dog only has to take a small step at first. Click and reward him for reaching the mat and feed him in position.
8. If you have moved at least a body length away from your dog and he still steps back to the mat with determination and fluency, you can introduce a verbal cue.
9. Continue to increase the distance between the chair and the mat, and prepare to fade the chair by sitting more and more on its edge so you are ready to stand up when you reach that point in training.
10. If the dog is still working well after two to three repetitions with the target mat, remove it for one or two attempts, and then put it down again. In this way you slowly work your way forward while the dog walks backwards through the alley.

Note: If your dog chooses to go forward instead of stepping back, you have probably put the treat at too great a distance from him. This can be remedied by blocking the front or side exit with a cushion, for example. Feed the dog when his back feet are on the mat.

Note: Your verbal cue for this exercise should be completely different from your cue for independent walking backwards, as in freestyle, for example. You will use this new verbal cue only when you are working in the alley and in combination with the dog's alignment to you.

Here you can see the sequence of movement that is required as the dog backs down an alleyway. The mat serves as a target for the back feet.

Equipment: Ikea stool (large), two cavaletti sets, one floor target.

FLEX +

CO-ORD +++

TACTILE ++

Exercise Band 1

2. Backing on to a sloping board (Advanced)

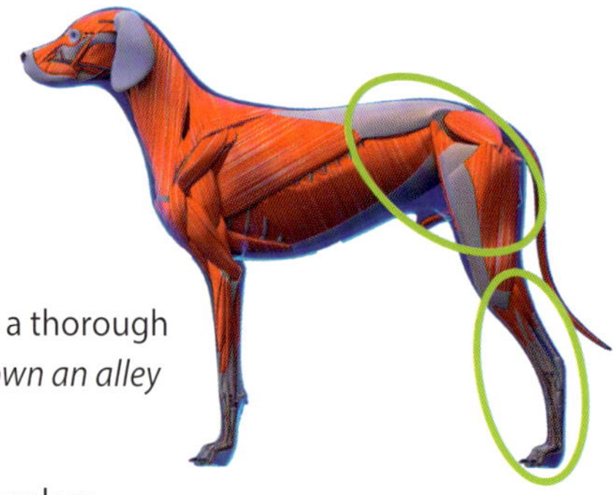

Do not attempt this exercise until your dog has a thorough understanding of the basic exercise, *backing down an alley* (see page 172).

All you have to do for the advanced exercise is replace the floor target with a sloping board. Make sure this is secured so that it does not slip. If necessary, set up the alley so the sloping board is positioned against a wall.

Click and reward the first few times the dog makes contact with the sloping board. This is a new training component and should, therefore, be rewarded with high-value treats.

Instead of using a mat as a rear target, the dog now aims for a sloping board.

Equipment: Four cavaletti sets, one sloping board.

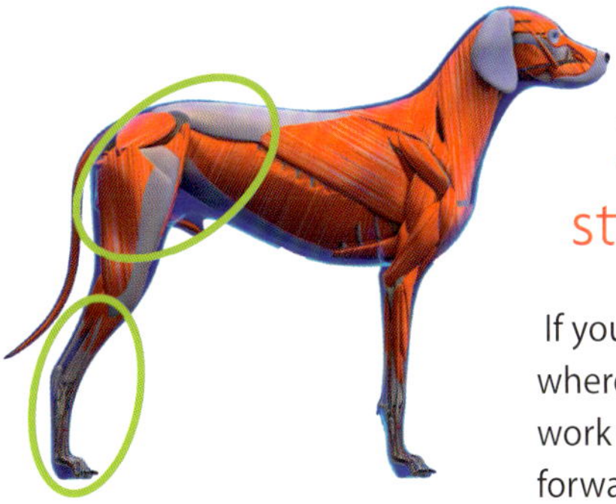

Exercise Band 1

3. Backing through an 'L' to stairs (Professional)

+ FLEX

+++ CO-ORD

++ TACTILE

If your dog has understood the previous exercises, where he has to back down an alley, you can now work with stairs at the end and then extend the alley forward.

1. Replace the sloping board, or mat, with a stair-like construction. Allow the dog to familiarise himself with the set-up, investigating it and walking backwards or forwards through it. Now position the dog a short distance form the stairs and encourage him to walk back to them.
2. Click the moment he places a back foot on the stairs, and reward generously. Repeat a few times.
3. Each time, start a little further away from the stairs until you reach the end of the straight alleyway. Then take a short break.
4. Allow the dog to relax while you prepare the new set-up. You will need an additional two cavaletti sets, which you should place at a 90 degree angle, in front of the existing alleyway to create an 'L' shape. In the initial stages, it will help if you the widen the angle so the dog has more room to manoeuvre.
5. When you start work again, ask the dog to back down the straight part of the alley. When he reaches the stairs, click and reward.
6. Now work gradually around the corner of the alley. First stand a little to the side and send the dog from there. Pay attention to your own position, as this will help the dog to align himself. If necessary, stand outside the 'L' to give yourself more space. Click and reward each step as the dog reverses round the corner.

Here you can see the construction of the 'L' with a mat as the target.

7. For the next step, stand a little more to the side and start the dog from this new position. Mark and reward the moment he turns his hindquarters around the angle.

8. Keep working – marking and rewarding – until the dog has got around the corner.

9. Now put the cavaletti sets closer together so that they form an 'L' shape. Do this incrementally and reward the intermediate steps.

10. If the dog is successful in negotiating the 'L', you can progress, step-by-step, until the dog starts at the beginning of the alley and, reverses all the way to the stairs.

11. In further training there are no limits to your creativity – you can extend this course to your heart's content! However, make sure you swap sides so the dog negotiates the turn using the other side of his body. Again, this must be trained gradually. Have fun!

Tip: If your dog finds it difficult to manoeuvre around the corner, check your own body alignment; perhaps you are turning too late so the dog does not have enough time to get around the corner? If you still can't get ahead, ask for the elephant trick (see page 35) in this set-up. Simply place the training equipment that you would otherwise use for the elephant trick within the angle. Now ask for the elephant trick and reward your dog generously for performing it. If this works, you can also perform the trick, within the angle, without any aids.

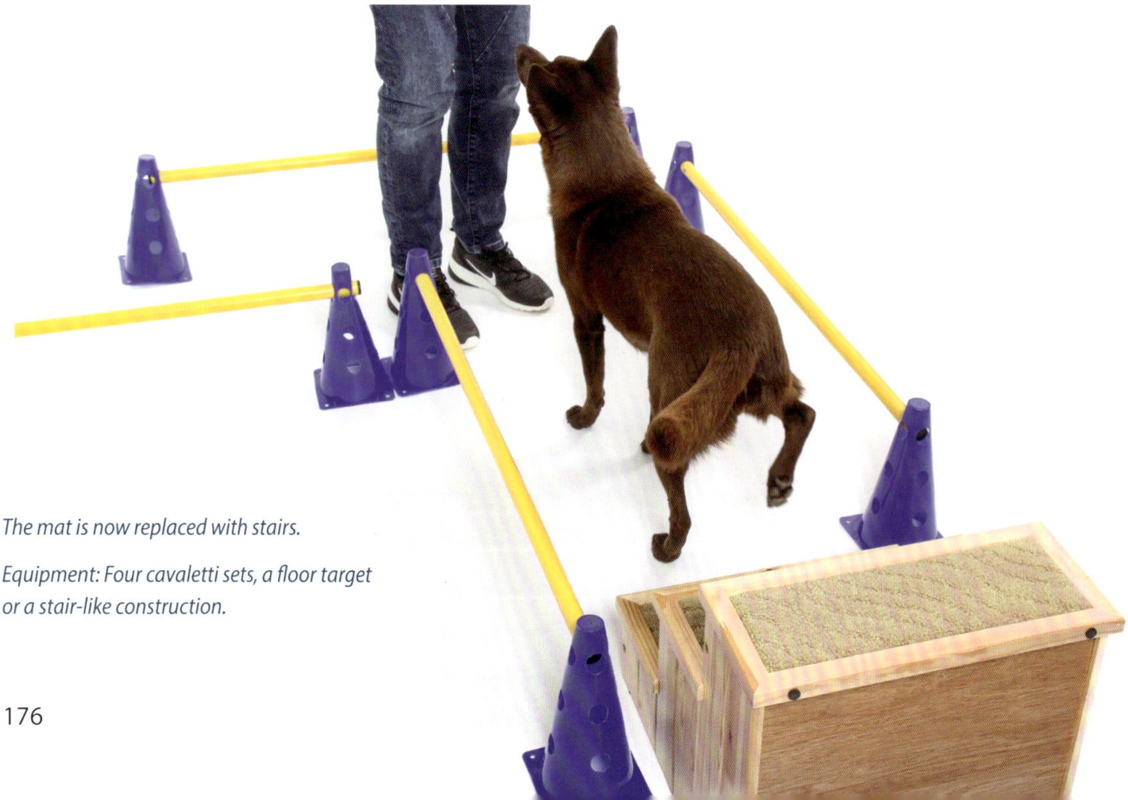

The mat is now replaced with stairs.

Equipment: Four cavaletti sets, a floor target or a stair-like construction.

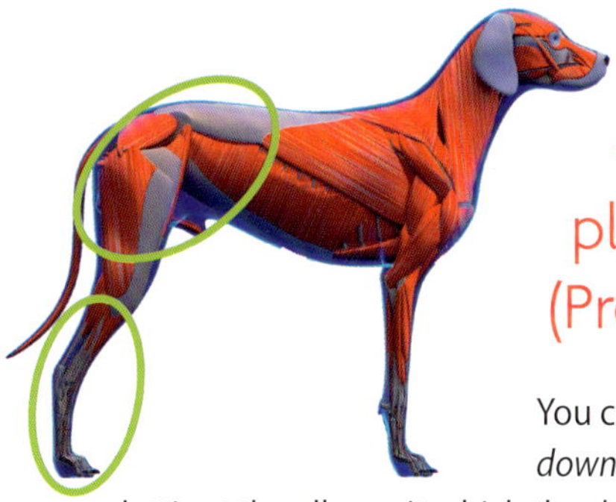

Exercise Band 1

4. Backing down an alley, plus side-stepping poles (Professional)

++ FLEX

+++ CO-ORD

+ STRENGTH

You can also progress the basic exercise, *backing down an alley* (see page 172), by placing more cavalettis at the alley exit which the dog must negotiate laterally. However, it is essential that you train this separately from the exercise for the fore limbs, *lateral cavaletti* (see page 73).

If the dog steps over the cavaletti sideways, you can then build an alley of cavaletti in front (see photo), and work on a fluid coordination work-out at the highest level which you can continue to build almost indefinitely. However, it is important not overdo it, so keep an eye on your dog for signs of fatigue.

This is the set-up for a backwards-sideways combination.

Equipment: Six cavaletti sets.

FLEX +

CO-ORD ++

STRENGTH +

Exercise Band 2

1. Backwards-forwards over a pole (Basic)

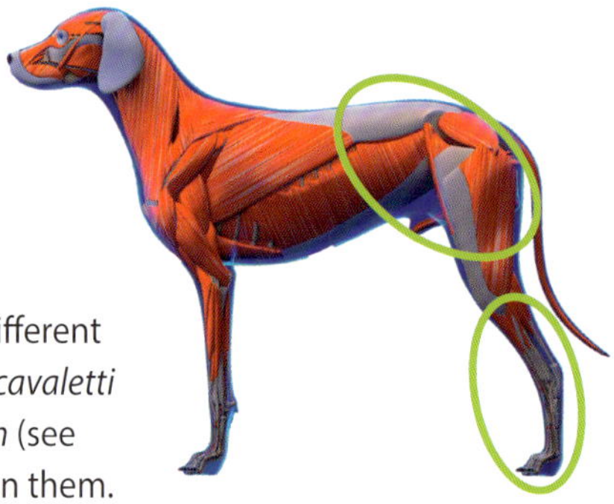

Make sure the set-up for this exercise is clearly different from the two exercises for the fore limbs, *lateral cavaletti* (see page 73), and *lateral cavaletti on an elevation* (see page 74) so that the dog can distinguish between them. You can do this by choosing different locations – for example, inside for the fore limb exercises, and outside for this one. You could also work with different equipment. You could train this exercise with a jump wings and a pole, and the others with a dog-stepper and a pole.

1. Your dog should know how to walk backwards on to a floor target (see *backing down an alley* (see page 172), but without the alley). Practise this a few times.

2. Now place a pole directly in front of the floor target and lead your dog over it, backwards, so he becomes aware of it.

3. Start the dog directly in front of the pole and give him the signal to walk backwards. You will need your clicker and treats at the ready because it's all about marking the moment the dog puts one back leg over the bar.

The dog needs to take a deliberate step over the pole, without outside assistance.

Equipment: Dog-Stepper.

4. The forward movement is given to you as a 'gift', which you can achieve through clever feeding. Do not feed the dog as he places his back foot over the pole but, instead, offer him the treat just in front of his muzzle. This will encourage him to reach forward and, thereby, bring his back foot over the pole again.

5. In most cases, the dog will always use the same leg when he is asked to step back over the pole. This could be because the dog favours one side – dogs are right- or left-handed – but it could be an indication that he is protecting the other leg. To test this, see if you can encourage him to use the other leg.

6. If you want to introduce a name for the behaviour, you will need to use separate verbal cues for the right and the left leg.

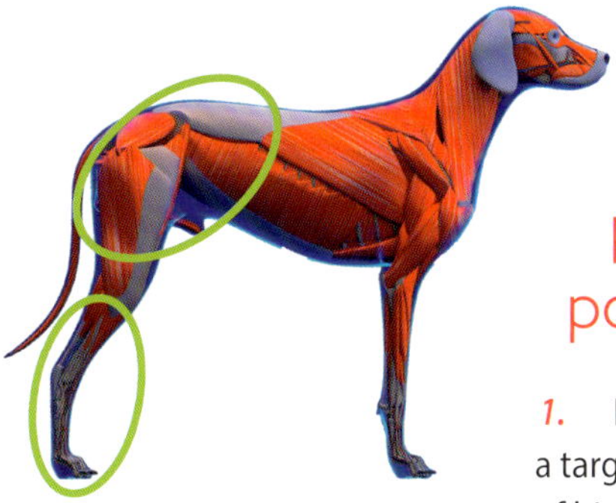

Exercise Band 3

1. Finding a back foot platform and holding position (Basic)

+ FLEX

++ CO-ORD

+ TACTILE

1. For this exercise, you will be using a chair and a target mat. In addition you will need a mixture of high-value and low-value treats – for example, cheese and kibble.

Place the mat in front of your chair in such a way that the distance is approximately equal to the dog's body length. When the dog stands directly in front of you, he should automatically place his back feet on the mat.

2. Now sit on the chair and lure the dog directly in front of you. Readjust the mat if necessary.
3. Place a low-value treat under the chair so that the dog has to step off the mat, with both back feet, to reach it. Now it gets exciting...
4. With your clicker and high-value treats at the ready, mark the moment when the dog moves backwards – after getting his treat – and steps back on the mat.
5. Repeat this procedure at least 10 times.
6. Introduce a verbal cue.
7. Now start placing a slight elevation behind the dog, ensuring it can accommodate his back feet. You can use a dog-stepper for this purpose. Make sure the elevation is secured and cannot slip.

8. Ask the dog to walk backwards again. Initially, he may be puzzled by the 'obstacle' so it is important that you click him immediately for touching it, and then reward.
9. Repeat several times, but delay the clicks until the dog places at least one back foot on the elevation.
10. Continue working until he stands with both back feet on the elevation.
11. Now click and reward him for staying in this position.

Note: If your dog steps forward, you have probably placed the treat at too great a distance. This can be remedied by blocking the front or side exit with a cushion. Feed your dog when his back feet are on the mat.

When the dog becomes proficient, you can extend the distance from the dog-stepper.

Equipment: Dog-Stepper.

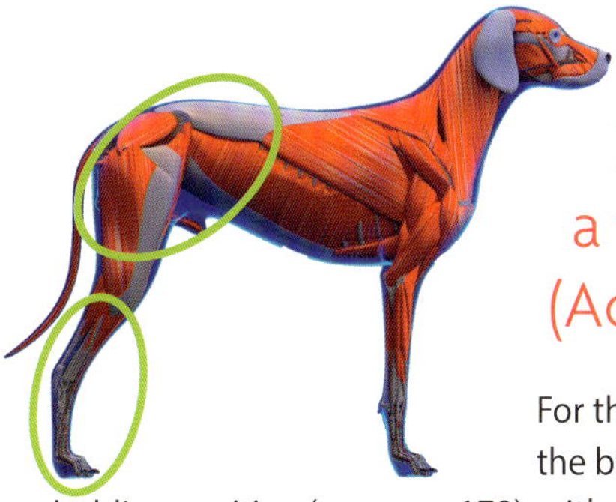

Exercise Band 3

2. Sit-stand transition on a wobbly back foot target (Advanced)

+++ CO-ORD

++ STRENGTH

+ TACTILE

For this exercise replace the stable elevation from the basic exercise, finding the back foot platform and holding position (see page 179), with a wobbly surface, e.g. a ball cushion.

1. Work with the dog, clicking and rewarding, until he is entirely comfortable with placing his back feet on a wobbly surface.
2. Now ask him to sit. Make sure you reward his initial attempts because sitting on a wobbly surface is quite a challenge.
3. When the dog is happy with sitting on a wobbly surface, do not reward this position but, instead, encourage him to stand and then feed him. He is then in the starting position ready to repeat the exercise.
4. Do not repeat this transition too often at first, as the dog will find it very tiring.

Tip: If your dog is reluctant to sit, try a larger target or practise sitting on a stable elevation first.

The starting position for the sit-stand transition is for the dog to stand with his back feet on the target, ensuring his hindquarters are parallel.

Equipment: TOGU ball cushion.

FLEX +

CO-ORD +++

TACTILE +

Exercise Band 3

3. Back feet on paw pods (Basic) and front foot change (Professional)

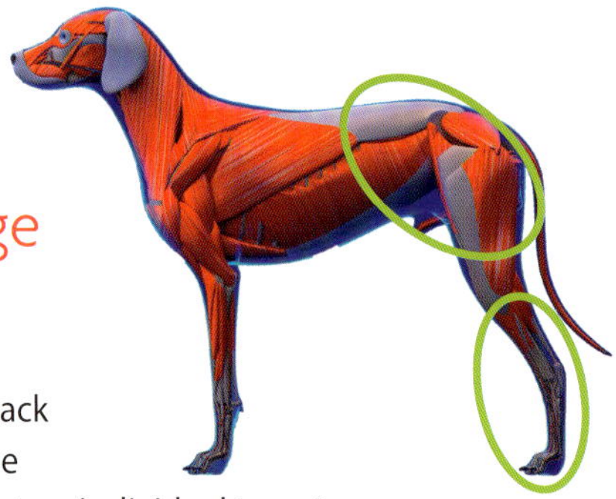

1. First refresh the basic exercise, finding the back foot platform and holding position (see page 179). Then replace the backhand target with two individual targets: for example, two paw pods.
2. Click and reward the dog for staying on the targets with his back feet.
3. When he becomes proficient at holding this position, you can start to work with his front feet.
4. For this purpose, sit down and position the dog in front of you, in a stand. Hide a treat in your fist and offer it to him level with his front feet.
5. Click and reward for anything that involves front foot action – do not click a nose touch.
6. Reinforce the front foot action with multiple clicks and rewards.

Tip: If your dog struggles with co-ordination, help him out, in the initial stages, by using larger targets for the back feet (for example, two ball cushions), and then gradually reduce in size.

The dog should be able to position his back feet, before you progress to working with his front feet.

Equipment: TOGU half balls.

7. Now turn your fist so that the back of your hand is facing down and reward for a touch.

8. Now offer an 'empty' fist; click and reward for a touch. Make sure you feed from the other hand.

9. Just before the dog starts his next 'paw stroke', open your fist so that he touches the palm of your hand.

10. Click and reward this several times, and then introduce a verbal cue for this foot. For the right foot you could use, "tip", since it contains an "i" just like "right". You can then call the left paw "tapp".

11. Now work with the other foot.

12. Do not always alternate between the front feet as it becomes highly predictable. Instead, challenge the dog, both physically and mentally, to listen out for the appropriate cue.

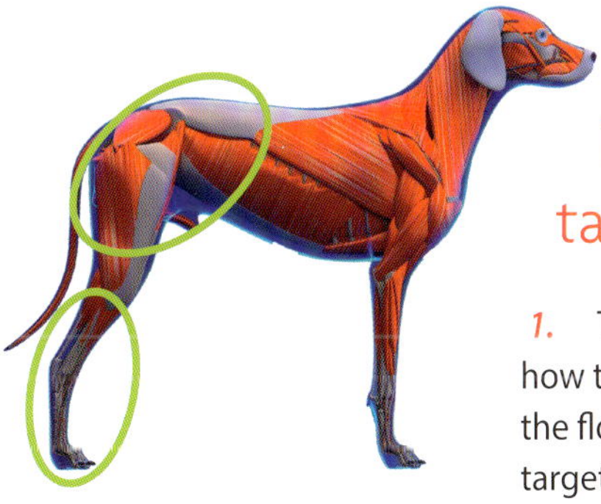

Exercise Band 4

1. Placing a back foot on a target (Basic)

+ FLEX

++ CO-ORD

++ TACTILE

1. This exercise is easier to teach if the dog knows how to walk backwards. First, place a large target on the floor in front of you. Walk him backwards over the target in such a way that there is little space between you, your dog and the target. Now get a chair and sit directly in front of the dog, who should be standing in front of the target. It is important that he stands, as no backward movement is possible if he is sitting or lying down. You should also ensure that he is not yet standing on the target.

2. Now it's getting exciting. Because you are sitting in front of the dog (but not touching him), and because the exercise set-up is similar to walking backwards, he is likely to offer this. Be quick and click the first movement of his back foot, and then work, incrementally, until he steps back on to the target.

3. Click and reward him several times for placing one back foot on the target.

4. With the next click, offer the food in such a way that the dog steps forward from the target, and is, thereby, ready to start the exercise again.

5. Generalise this exercise with a variety of targets, especially smaller ones so the dog becomes skilled at placing his foot. At this stage you can introduce a verbal cue.

6. You are now ready to start work with the other back foot.

Generalise the exercise with a variety of targets.

CO-ORD ++

STRENGTH +++

TACTILE +

Exercise Band 4

2. Lifting alternate back feet (Advanced)

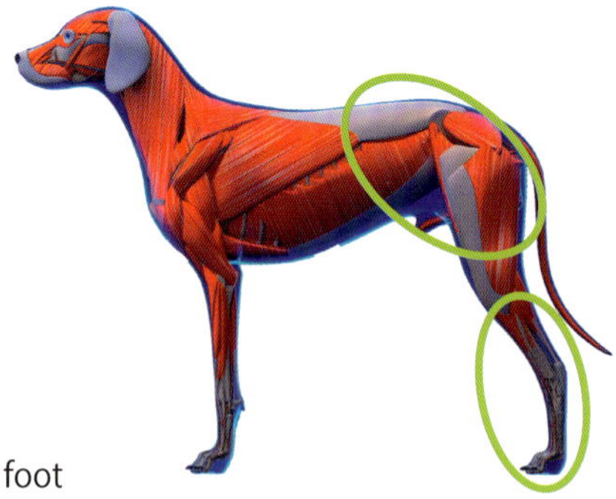

When you have taught placing a back foot on a target (see page 183), you can shape him to lift his feet.

1. Refresh the basic exercise of placing a back foot on a target.

2. Now your clicker skills come into play as this next step is all about timing. You must now mark the moment the dogs lifts his back foot to place it on the target.

3. Click and reward this several times.
4. Keep practising and when you are successful at least 15 times in a row, you can introduce a verbal cue. The next step is to remove the target and ask for the behaviour solely on a verbal cue.
5. Now start work on the other back leg, following exactly the same procedure.
6. When the dog fully understands the exercise, you can ask him to lift alternate feet. This has a great strengthening effect, as well as improving co-ordination.

Lifting alternate back feet can be shaped by training the dog to step on targets.

Equipment: TOGU half balls.

Tip: If you are struggling to mark the behaviour, you can increase the height of the target so the dog has to make a more obvious movement in order to reach it.

When the dog understands the exercise, he will be able to lift his back feet on a verbal cue, without needing targets.

FLEX +

CO-ORD +++

STRENGTH ++

Exercise Band 4

3. Lifting alternate back feet/wobbly (Professional)

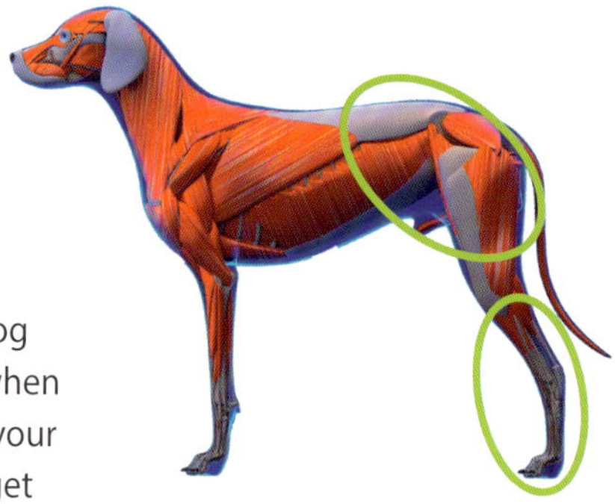

When you start work on this exercise, give the dog time to find his balance on the wobbly surface when he is standing. When he is happy to do this, use your verbal cue to lift a back foot, offering a hand target to make it easier for him. This exercise demands balance and co-ordination, so make sure you do not over-tire your dog.

Lifting a back foot when standing on a wobbly surface is quite a challenge. Offering a target (in this case a hand) will facilitate learning.

Equipment: TOGU Balance Ballstep.

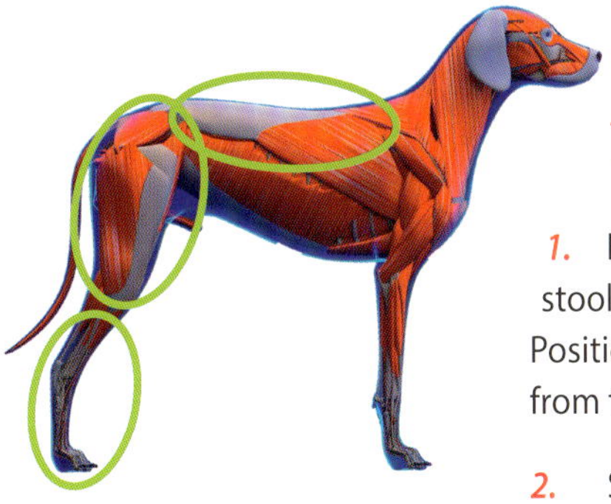

Exercise Band 5

1. Reverse slalom (Basic)

++ FLEX

+++ CO-ORD

+ STRENGTH

1. For this exercise, you need to use a chair or a stool, from which the dog can eat when standing up. Position him at a distance of approximately 2m (6ft) from the stool.

2. Stand directly in front of the feeding station so that your toes touch it. With your clicker and treats at the ready, place a treat on the stool.

3. Now call the dog through your legs to the feeding station, where he will find his treat.

4. The dog will want to exit after feeding, and the easiest option is reverse through your legs. Wait for this moment, and click as he goes through your legs. Feed from behind, making sure you keep the dog in a straight line.

5. Once again you must play the part of a treat-vending machine! The treat on the stool is for position only; it is the treat behind your back that is key to teaching the behaviour. In this scenario, it is a good idea to work with treats of different value. For example, use a low-value treat, such as kibble, on the stool, and a high value treat, such as cheese, when you are rewarding from behind. Keep working on step 4, with the dog going forward to get the treat and then reversing through your legs until it becomes a fluid movement.

6. Once you have established this sequence, gradually increase your distance from the feeding station. Make sure you also create a distance between yourself and the dog before you start. This prevents him orienting himself towards your body and anticipating the movement. So, ask the dog to sit, move away from the stool in a straight line, stop, put down a positioning treat and start the exercise.

Using the small-step exercise structure, the dog learns how to weave through each leg independently.

Note: If your dog tries to step forward, you can place cushions, or similar, around the stool, so that it is impossible for him to slip through to the front or to the side.

7. Continue to move away from the stool until you reach the point where the dog's hips are between your legs – the rest of his body 'sticks out' facing the stool. Now the process of turning in begins. At the moment you are still standing hip-wide in front of the feeding station. Imagine that one of your legs is moving a quarter circle, backwards. It doesn't matter which leg you choose, but it should always be the same one. Let us start with the right leg. Position your right leg on the imaginary quarter circle, a little to the rear, making sure you are only slightly offset. Position yourself in such a way that when the dog walks backwards between your legs, he will bump against your leg. This moment is important: click as he bumps your leg and reward. Repeat this step several times.

8. Now take one step further back with your right leg and repeat. Keep working on this and, if the dog is successful, keep moving your right leg back until it is in line with your left leg. Proceed in easy stages, rewarding the intermediate steps frequently.

9. If the dog is able to reverse through your legs, from the feeding station, and then bend, introduce a verbal cue. For example, you could use as "zick" for the right side, and "zack" when you teach the left side. It is also helpful if you reward from your stationary leg, i.e. if you move your right leg, reward from the left leg.

10. You have now built up one side of the reverse slalom and placed it on a verbal cue. Don't worry, the other side is easy!

11. Go back to the beginning and, this time, move your left leg in an imaginary quarter-circle and link the movement with a verbal cue, e.g. "zack".

12. Now you can put the back slalom together. The dog needs to be next to you in the heelwork position, either on the right or left side, depending on which leg you plan to move first. Forget the leg sequence you have used up until now; this is what to do next. First move your outside leg back (the one furthest from the dog). As you do this, give your verbal cue for the corresponding side. If you have set your right leg backwards first, and assigned the cue "zick" to this exercise, your dog will have linked a left bend with it. If this goes well, reward generously.

13. Now try the other side. If this also works, you can put both sides together and go for the complete exercise – a backwards slalom!

Option:

Work towards linking both backward bends with a verbal cue so the dog doesn't need to do the complete slalom. You can train a specific bend individually. For example, a bend to the right is trained by having the dog sit on your left and asking him to weave backwards through your legs, with your right leg stepping backwards. If you call this side "zig" and the other side "zag", you can work on them as separate exercises.

X. SPECIFIC NEEDS

Health Issues

There are many orthopaedic and neurological conditions that require a training plan but most are either too complex, or belong in the hands of a professional therapist from the outset. However, there are three conditions – hip dysplasia, patella luxation and cauda equina syndrome – which can benefit from a basic training programme. This is always with the proviso that you seek expert advice, and your dog has a thorough assessment before attempting any of these exercises.

HIP DYSPLASIA (HD)

Hip dysplasia is a very complex disease which, in the course of time, affects not only the hip but extends over the entire musculoskeletal system. It is a predominantly an hereditary disease of the hip joint, in which instability of the joint occurs due to a mismatch between the head of the joint and the socket.

The hip joint plays a very important role in locomotion, because the drive for forward movement is mainly provided by the hip joint. So if a young dog experiences pathological movement stimuli with every step, pathological ossification stimuli, and secondary changes due to relieving posture patterns in the entire musculoskeletal system, will naturally occur. From our point of view, there is no such thing as a minor case of HD.

The knee joints (X-B adjustment), the transition from the sacrum to the lumbar spine (cauda-equina compression syndrome), the transition from the lumbar spine to the thoracic spine (increased shear forces), the elbow joints (often diagonally to the bad side) are very often overloaded, and the cervical spine (compensatory pendulum movement) is also affected. From around six to eight years of age, depending on breed and size, there is also a restriction of movement in the space of the forelimbs due to insufficient movement of the shoulder blades, muscle tension, trampling of the metacarpals, nerve irritation/nerve glide deficits.

Our goal with the growing dog is for an orthograde weight and swing distribution, and to harmonise the gait pattern. If the problem is still evident in a growing dog, there is the possibility of achieving an improved shape by optimising the pressure of the femoral head in the growing joint socket through targeted training and, if necessary, supportive pain therapy.

With the middle-aged dog, it is important to stop the 'overload construction sites' and to optimise the muscle status.

In older animals with HD it makes sense to keep the entire shoulder area mobile, as well as the cervical spine, and to plan a variety of relaxation exercises for the back.

The pain tolerance limit is usually very high, as these dogs have to deal with this disease from a young age, so it is important to have the current pain status clarified by a vet. A dog that has to tolerate pain not only suffers stress, he is also unable to build up muscle.

Training plan

To relax the tensed inner thigh muscles and improve the outer thigh contour, see:
V. Exercises for the outer thigh muscles
Exercise Band 1: Exercise 1 – stepping sideways over a pole (page 119)
Exercise Band 2: Exercise 1 – elephant trick (page 35)
Exercise Band 4: Isometric exercises (page 126)
Exercise Band 5: Exercise 1 – side stepping with front and rear limbs stationed (page 130)

To improve body awareness and co-ordination of the hindquarters, see:
IX. Exercises to train body awareness and co-ordination of the hind limbs
Exercise Band 1: Exercise 1 – backing down an alley (page 172)
Exercise Band 4: Exercise 1 – placing a back foot on a target (page 183)

To improve mobility, see:
- Cavaletti training (see page 204). The height of the cavaletti should be no higher than your dog's hock.

To improve back relaxation, see:
VI. Exercises to relax the back
Exercise Band 1: Exercise 1 – plank (page 134)
Exercise Band 5: Exercise 1 – look under the belly (page 144)

For relaxing the front, shifting weight to the rear and working in the opposite direction of the cervical spine, see:
II. Exercises for the neck and shoulder muscles
Exercise Band 3: Exercise 1 – elevated stand (page 56)

To improve step length of the front limbs, see:
III. Exercises for the fore limbs
Exercise Band 8: Exercise 1 – from paw to paw (page 86)
Exercise Band 10: Exercise 1 – front leg forward (page 90)

The degree of HD suffered by an individual dog depends on fitness and age. It is, therefore, advisable to seek the help of a physiotherapist in drawing up a training plan and adjusting it when necessary.

PATELLA LUXATION

Patella luxation is a displacement of the patella (usually inwards) from its natural position. Very often this disease occurs in small breeds and rarely is it traumatic (i.e. the result of an accident).

The causes in the congenital form include an excessively flat sliding groove, axial deviation of hind limbs, displaced quadriceps group, and muscle atrophy of the quadriceps muscles. There may be a short-term occurrence when a bitch is in season, or as a result of poor connective tissue due to producing too many litters.

Affected dogs show a sudden lameness, which is expressed by holding the affected leg up for a few steps (flexion of knee and ankle). While holding it up, the kneecap is outside the glide channel, and when the kneecap slides back, the leg is loaded normally again.

Patella dislocation is divided into four grades, whereby conservative therapy can be attempted for grades one and two. However, surgery is recommended for grades 3 and 4 in order to avoid permanent damage to the articular cartilage and inflammation in the joint and surrounding structures.

Our aim is to achieve a good muscle build-up and, thus, provide better stability of the joint, which should prevent the kneecap from slipping out. The following exercises should only be attempted after consultation with a veterinarian and/or physiotherapist.

Training plan

For strengthening the quadriceps group see:
IV. Exercises for the front thigh muscles
Exercise Band 1: Exercise 1 – sit-stand-sit (page 95),
Exercise 2 – sit-stand-sit/front elevated (page 96)
Exercise Band 2: Exercise 2 – lifting the buttocks (page 103)
Exercise Band 3: Exercise 1 – narrow cavaletti (page 105)
Exercise Band 4: Exercise 1 – climbing stairs (page 106)
Exercise Band 5: Exercise 1 – lunge forward (page 108)
Exercise Band 6: Exercise 1– weight shift on a wobble board (page 110)
Exercise Band 7: Exercise 1 – bounce to a hand touch (page 111)
Exercise Band 8: Exercise 1 – mountain goat (page 113)

In the case of frequent luxation, back tension can also occur and a relaxation programme for the back, as described in **V1. Exercises to relax the back muscles** (page 132) can be applied afterwards.

During training, care should be taken to ensure that no dislocation occurs; it may be useful to have an assistant stabilise the knee.

CAUDA EQUINA SYNDROME

This clinical condition is usually found in large dogs but medium-sized dogs, who take part in competitive sports, often show problems in the lumbosacral transition.

Cauda equina syndrome, also known as degenerative lumbosacral stenosis, can have various causes and usually develops insidiously (except for trauma or infection). This results in a narrowing of the transition between the lumbar spine and the sacral vertebrae; the spinal cord tapers at this point to nerve endings resembling a horse's tail (hence cauda equina). Often, the cauda equina-forming nerves swell up, resulting in a narrowing of the spinal canal.

The first symptoms are seen as measuring jumps (car, sofa, stairs, sports), sudden change in behaviour due to pain, problems getting up, touch sensitivity at the rear of the back, and a dragging gait. With advanced disease progression, incontinence and dowsing paralysis can also occur.

Common causes:

- *Degeneration/wear due to overloading and incorrect loading. These include external factors, such as sports overloading, but also poor hip placement or persistent lameness due to torn cruciate ligaments, etc.*
- *Blocking ISG/SIG (Blocked or displaced ileosacral joint) resulting in a contracture of the lumbar musculature, with premature ageing of the intervertebral disc between L7 and S1, irritation of the cauda equina and overstretching of the ventral ligament below the spine (resulting in osteophyte formation).*
- *Trauma.*
- *New formation/tumour (rare).*
- *Infection (rare).*
- *Genetic (transition vertebrae, step formation, etc.).*
- *More males are affected, which may be due to a segmental reflective reaction of the structures by a change in the prostate.*

In many cases surgery is not necessary, but pain medication will be required. Nevertheless, you should try an adapted training programme in an attempt to preserve the muscles of the hind limbs (except the quadriceps group), and to improve nerve transmission. Following surgery, training can be started after the rehabilitation phase, but this must be in consultation with your veterinarian or physiotherapist.

Training plan

IX. Exercises to train body awareness and co-ordination of the hind limbs
Exercise Band 1: Exercise 1 – backing down an alley (page 172)
Exercise Band 3: Exercise 1 – finding a back foot platform and holding position (page 179)
Exercise Band 4: Exercise 1 – placing a back foot on a target – focus on worst side (page 183)

VI. Exercises for the outer thigh muscles
Exercise Band 2: Exercise 1 – elephant trick – focus on worst side (page 35)
Exercise Band 5: Exercise 1 – side-stepping with front and rear limbs stationed – focus on worst side (page 130)

V1. Exercises to relax the back muscles
Exercise Band 1: Exercise 1 – plank (page 134)
Exercise Band 3: Exercise 1 – stretching (page 141)
Exercise Band 5: Exercise 1 – looking under the belly (page 144)

Seniors

What classifies a dog as senior depends on various factors, such as genetics, breed (size), lifestyle (e.g. stress), health (diseases) and, last but not least, nutritional status. In larger dogs, ageing begins as early as six years of age; giant breeds are affected earlier, small breeds around eight years of age.

Biological ageing is defined by a degenerative biological process which results in physical and mental deterioration. Chronic diseases often lead to an earlier onset of ageing, as the body loses its ability to compensate through pain-relieving posture patterns, and the resulting permanent stress on the body, drives the aging process forward.

In addition to the external and, thus, visible changes, such as the appearance of white hair around the mouth, dwindling of muscle mass and performance, there are also changes in oxygen metabolism, due to reduced cardiac output and changes in the vascular system. Furthermore, the ageing process also leads to metabolic changes, i.e. reduced metabolism and diminishing performance of the immune system.

With increasing age there is also a reduced sensory performance, gait insecurity, loss of co-ordination and dementia-like symptoms, which can make everyday life with a senior citizen considerably more difficult. In physiotherapy, we often experience older patients who, like older people, find it difficult to cope with deviations from everyday life and are easily stressed. Often there are reports of nightly 'urge walks' and restlessness.

At Saarland University Hospital in Homburg, Germany, researchers investigated how regular exercise affects cell ageing in humans. The research focuses on the interior of the nucleus of white blood cells (leukocytes). This is where the chromosomes are stored which contain genetic information (DNA). Each chromosome has so-called telomeres at the end of its DNA strands. They protect the DNA, similar to the caps at the end of shoelaces. However, with each cell division, the protective telomeres shorten, so that the natural ageing of the cell progresses. According to the researchers, the telomeres can be 'lengthened' through sport. Since the physiological processes in the body are almost identical in humans and mammals, we can assume the same effect occurs in dogs.

Therefore, in addition to a fitness programme, which is within the individual dog's capabilities, we should, also include elements from a sensory programme, along with cognitive stimulation. The effect of this type of programme is clear to see and we find it is particularly rewarding to enrich the lives of our older dogs.

Focus for senior training:

- *Joint mobility in general*
- *Improving mobility in the chest and shoulder area*
- *Muscle maintenance*
- *Improving body awareness*
- *Relaxation of overloaded structures (often back and shoulder area)*
- *Cognitive occupation*

Warm-up programme

The warm-up programme should be is adapted to the individual dog. It is important not to over-tire your senior, so focus on gentle preparation for exercise. The duration of the training depends on the dog's fitness and, sometimes, how he feels on the day. Don't be fooled by your dog's enthusiasm! If you see signs of stiffness after the warm-up, you have, clearly, done too much.

Exercises

1. Cavaletti training:

Stepping over 4-5 cavaletti poles: The height should be adjusted to correspond to the distance between the bend of the toes and the hock; the distance between poles should equate to the dog's height at the withers.

The dog should slowly, and deliberately step over each pole and, under no circumstances, jump! Treats usually have a negative effect on cavaletti training, as the dog's focus is more on the food than on foot placement. Therefore, it is a good idea to build up the cavaletti training using floor targets. This is important because seniors are often more food-oriented than younger athletes, and, in addition, they no longer have such good co-ordination.

2. Exercises to improve the mobility of the cervical spine and stretch shortened muscles in the shoulder and forelimb area, see:

II. Exercises for the neck and shoulder muscles:
Exercise Band 1: Exercise 1 – nose to chest (page 50)
Exercise Band 2: Exercise 1 – stretched sphinx (page 53)
Exercise Band 3: Exercise 1 – elevated stand (page 56)
Exercise Band 4: Exercise 1 – head to shoulder (page 59)

3. Relaxation of the back muscles, see:

V1. Exercises to relax the back muscles
Exercise Band 1: Exercise 1 – plank (page 134) Exercise Band 4: Exercise 1 –knee/flank stretching (page 142)
Exercise Band 5: Exercise 1 – looking under the belly (page 144)

4. Exercises to improve shoulder mobility and loosen adhesions in the trunk area, see:

III. Exercises for the fore limbs
Exercise Band 2: Exercise 1 – lateral cavaletti (page 73) Exercise Band 8: Exercise 1 – from paw to paw (page 86)
Exercise Band 10: Exercise 1 – front leg forward (page 90)

5. Exercises for building thigh muscles (as these muscles often deteriorate with age), see:

VI. Exercises for the outer thigh muscles
Exercise Band 2: Exercise 1 – elephant trick (page 35)
Exercise Band 4: Exercise 1 Isometry: lifting front legs (page 126), Exercise 2 – thigh pressure (page 127), Exercise 3 – lifting back legs (page 128)

6. Improvement of body awareness (especially hindquarters), see:

IX. Exercises to train body awareness and co-ordination of the hind limbs
Exercise Band 1: Exercise 1 – backing down an alley (page 172).
Exercise Band 2: Exercise 1 – backwards-forwards over a pole (page 178)
Exercise Band 3: Exercise 1 – finding a back foot and holding position (page 179)

For older dogs, a sensory track (tactile stimuli) is also recommended. It is easy to set up using a variety of different surfaces – just make sure you do not use anything that is too wobbly. The aim is for the dog to work slowly and calmly as he must consciously adjust to each new surface. After you have finished work, you can play a search game and/or give your dog a gentle massage to finish the session. Grateful seniors are guaranteed!

Here you can see examples of how to set up a sensory track.

Early motor development for puppies and young dogs

Puppies should be exposed to early, controlled, learning experiences. These will provide the foundation for later life and also for success in sports.

In older dogs, the nerve absorption and transmission of stimuli diminishes, but if the individual responds to stimuli early on, these structures will be developed and can be better retrieved in old age and illness.

Breed differences

We have such a wealth of different breeds, the training for a young dog will depend on his conformation and, of course, his temperament.

A Great Dane will soon be over-taxed with exercises that require good co-ordination, whereas a small terrier can accomplish the most amazing feats at 10 weeks of age. Good angulation of the joints allows for shock absorption, with the hindquarters providing thrust and power (particularly for jumping). If the shoulder blade is too steep, it will affect stride length, an issue that also occurs if the front legs are too far in front of the chest. This, in turn, exposes the dog to greater stress on landing. A balanced, harmonious physique is therefore a prerequisite for a good movement pattern.

In training, you should address all aspects as far as possible, but especially co-ordination, since good balance and responsiveness are essential for an active dog. Co-ordinative development is completed at the end of growth, so a dog will need to undergo strategic learning experiences by this time. So, if your dog has a great agility career ahead of him, he should start with sensible, measured foundation training.

Important training points:

- *Encourage corrective reactions in large dogs (e.g. steps).*
- *Fine motor skills/tactile stimuli (e.g. different surfaces).*
- *Assessment of ups and downs (climbing tasks).*
- *Body awareness (wearing chest harness, coat etc.).*
- *Practise untypical movement sequences (stairs, walking backwards).*
- *Maintain created movement patterns (creep and spinning).*
- *Playing and moving in shallow water and on deep surfaces (e.g. sand, leaves).*
- *Rhythm ability (cavaletti, jumps).*

As we have seen when looking at senior dogs, exercises for building thigh muscles are important because these muscles often deteriorate with age, usually resulting in a narrow, restricted gait.

Important: these exercises should be trained slowly, calmly, and in absolute moderation!

Training examples:

- *Rest station (different targets for stationing).*
- *A sensor system co-ordination path consisting of 3-4 different objects of varying height and nature.*
- *1-3 Cavaletti poles (pole height: toe bend to metacarpus).*
- *Learning to walk backwards to a rear target*
- *Climbing station.*
- *Body perception (touch, body targets, neck collar and coat training).*

Important: Controlled and targeted exercises are preferable to an environment where the puppy has free, unsupervised movement, such as in the garden.

Remember, a puppy will struggle to find his balance and should, therefore, not be exposed to extremely wobbly surfaces.

When can I start using the adult programme?

The sacrum usually fuses at the age of 18 months, which can be used as a marker for the completion of growth. Up to this age, the joints should not be subjected to undue stress.

Duration of training

A good guideline for puppies is to allocate one minute of training for each month of their age (or as long as 10-15 treats last). Different priorities should be addressed one to three times per week, but it is important to make sure that puppies and young dogs have days in between where no training takes place.

Joint diseases in growing dogs

If you detect any lameness in a growing dog, it must be taken seriously and a thorough clinical and orthopaedic examination should be considered essential.

Skeletal problems, accompanied by lameness, usually occur between the fifth and ninth month of life, as this is when the greatest growth spurt occurs. In addition to the risk of traumatic incidents due to insufficient motor experience, genetic predisposition and unbalanced nutrition (too much energy), young dogs are at their most active during this period. This, in turn, is often answered with too much unsuitable physical activity (e.g. romping with other dogs).

The most common diseases that occur during this period are hip dysplasia (HD), elbow dysplasia (ED), panostisitis, and osteochondrosis/osteochondritis.

An early diagnosis of HD between the sixteenth and twentieth week has not yet become widely accepted, as owners shy away from sedation at this age. In the case of at-risk breeds or crossbreeds involving predisposed breeds, early diagnosis at this age can make sense. This is because it is still possible to have an early surgical, or conservative, influence on the musculoskeletal system in order to minimise consequential damage. We would like to briefly mention the two of the currently best-known methods:

1. The anatomical-morphological assessment of the growing hip joint between the sixteenth and twentieth week of life, according to KÖPPEL, in which the possible degree of HD is concluded due to the ossification processes at the acetabulum.

2. The distraction index – "PennHIP®" after SMITH (modified), which is given as a measure of the looseness of the joint and attempts to derive the expected degree of HD from this.

XI. IN THE GYM

Circuit training for canine athletes

THE SET-UP

So far, we have described all the exercises in this book as separate units to work on, and perfect. But once your dog becomes proficient, it is fun to set up a sequence of exercises, with equipment, and have a go at circuit training.

If your dog has no physical issues, and is at peak fitness, you can be really creative in what you set up. However, it still makes sense to devise a set-up with a specific goal in mind.

By working through the exercises outlined in this book, it is evident that training rarely applies to a single muscle group. If exercises are carefully planned, and well executed, they can have a positive effect on different muscles and muscle groups. So, when we are planning a sequence of exercises, we need to give very careful consideration to what is happening in the body, and what we hope to achieve.

First and foremost, the dog needs to have mastered the individual exercises you have in mind; you can then plan a running order. It is a good idea to set up a station, where you can 'park' the dog in-between exercises if you need to make changes. Teach him a clear 'start' signal, and a signal that means pause, so you can call a halt and park him, if necessary.

Recommended running order

- *Warm-up*
- *Exercises for co-ordination and strength (always introduce directly after the warm up)*
- *Stationary exercises (focus on these after the dog has done something more strenuous)*
- *Targeting tricks (introduce further into the sequence)*
- *Tactile exercises (introduce further into the sequence)*
- *Relaxing the muscles (leave until the end of the sequence, otherwise the risk of injury is too high).*

IDEAS FOR SEQUENCE TRAINING

1. Response to tactile stimuli in motion – setting up targets.
2. Combine tactile stimuli with co-ordination.
3. Tactile stimuli with body targets – for nervous dogs.
4. Improve co-ordination in general.
5. Improve co-ordination of the forehand.
6. Improve hindquarters co-ordination.
7. Back relaxation.
8. Train back muscles.
9. Hindquarters in general.
10. Hindquarters outer thigh muscles.
11. Hindquarters front thigh muscles.
12. Hindquarters hip extensor.
13. Hindquarters hip extension plus tactile stimuli.
14. Cervical spine.
15. Improving spine flexibility.
16. Shoulder training.
17. Fore limbs.
18. Shoulder-torso-front limb.
19. Sideways track.
20. Reverse track.
21. Elephant trick combos.
22. Cavaletti variations.
23. Training for senior citizens.
24. Training for puppies.
25. Mixed circles

You see, there are no limits to your imagination!

Combination Exercises

1. Response to tactile stimuli in motion – approaching targets (little movement)

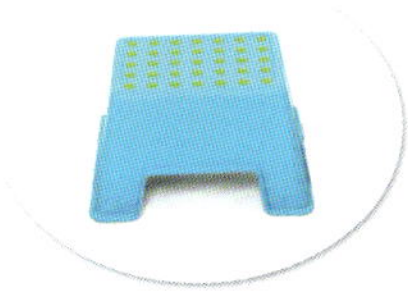

II. Exercises for the neck and shoulder muscles
Exercise Band 3: Exercise 1 – elevated stand/basic (page 56) with target for stationing.

III. Exercises for the fore limbs
Exercise Band 6: Exercise 1– looking down from an elevation/basic (page 80) – effect increased with paw pods.

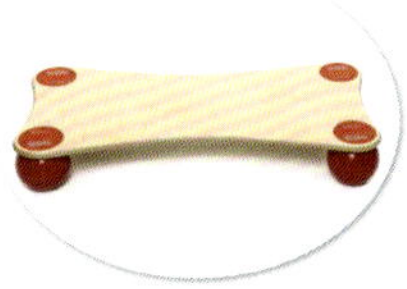

III. Exercises for the fore limbs
Exercise Band 1: Exercise 2 – isometry with hand on shoulder/wobbly/advanced (page 72).

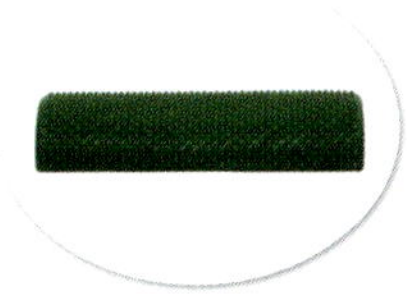

IX. Exercises to train body awareness and co-ordination of the hind limbs
Exercise Band 3: Isometric exercises, e.g. Exercise 1 – finding a back foot platform and holding position/basic (page 179).

2. Combine tactile stimuli with co-ordination

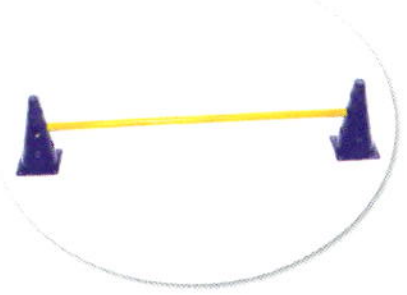

IX. Exercises to train body awareness and co-ordination of the hind limbs
Exercise Band 1: Exercise 1 – backing down an alley/basic (page 172).

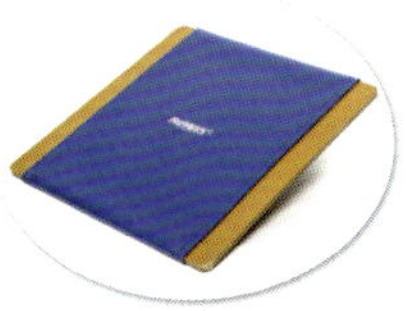

IV. Exercises for the front thigh muscles
Exercise Band 6: Exercise 1 – weight shift on a wobble board/basic (page 110).
Backwards over the wobble board three times and turn combined with:
VI. Exercises to activate the back muscles
Exercise Band 3: Exercise 2 – tight spin/wobbly /advanced (page 152)

IV. Exercises for the front thigh muscles
Exercise Band 1: Exercise 4 – knee bend on a peanut ball/professional (page 98), Exercise 5 – sit on a peanut ball/professional (page 100)

3. Tactile stimuli with body targets – for excitable dogs

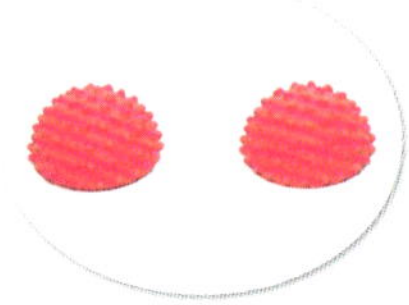

- **III. Exercises for the fore limbs**
  Exercise Band 5: Exercise 1 – pulling forward/basic (page 79).

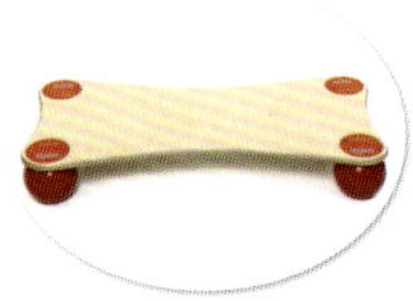

- **III. Exercises for the fore limbs**
  Exercise Band 1: Exercise 2 – isometry with hand on shoulder/wobbly/advanced (page 72).

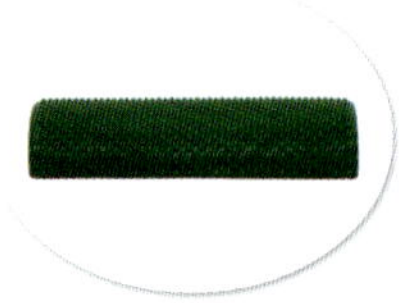

- **VI. Exercises to activate the back muscles**
  Exercise Band 1: Exercise 1 – backing to a target/basic (page 147).

- **VI. Exercises to activate the back muscles**
  Exercise Band 2: Exercise 1 – balancing act/basic (page 149) with:
  II. Exercise for the neck and shoulder muscles
  Exercise Band 5: Exercise 1 – hold/basic (page 63).

4. Improve general co-ordination

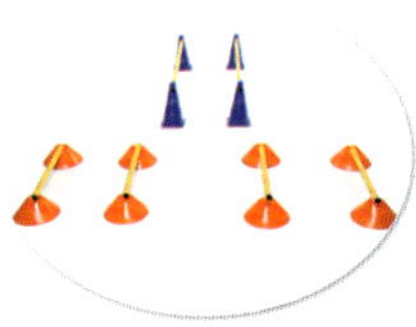

- **IX. Exercises to train body awareness and co-ordination of the hind limbs**
  Exercise Band 1: Exercise 4 – backing down an alley, plus side-stepping poles/professional (page 177).

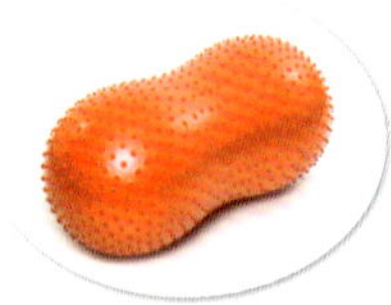

- **IV. Exercises for the front thigh muscles**
  Exercise Band 1: Exercise 5 – sit on a peanut ball/professional (page 100).
  IX. Exercises to train body awareness and co-ordination of the hind limbs
  Exercise Band 4: Exercise 3 – lifting alternate feet/wobbly/professional (page 186).

- **VI. Exercises to activate the back muscles**
  Exercise Band 3: Exercise 2 – tight spin/wobbly/advanced (page 152).

- **IV. Exercises for the front thigh muscles**
  Exercise Band 7: Exercise 2 – bounce to a hand touch/wobbly/advanced (page 112).

5. Improve forehand co-ordination

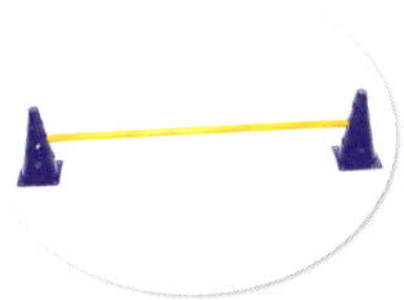

III. Exercises for the fore limbs
Exercise Band 2: Exercise 1 – lateral cavaletti/basic (page 73).

III. Exercises for the fore limbs
Exercise Band 3: Exercise 2 – lateral cavaletti on an elevation/advanced (page 74).

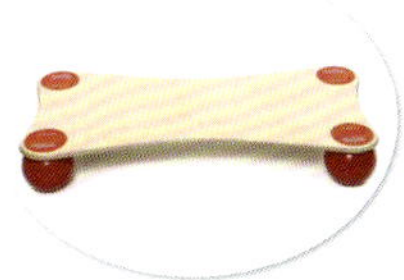

III. Exercises for the fore limbs
Exercise Band 8: Exercise 1 – from paw to paw/basic (page 86)

6. Improve hindquarters co-ordination

IX. Exercises to train body awareness and co-ordination of the hind limbs
Exercise Band 1: Exercise 3 – backing through an 'L' to stairs/professional (page 175).

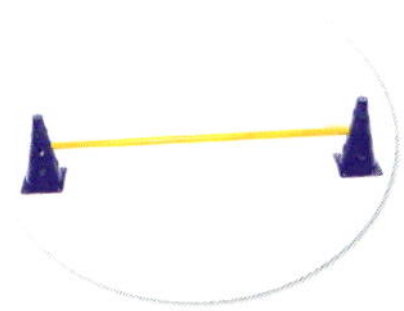

IX. Exercises to train body awareness and co-ordination of the hind limbs
Exercise Band 2: Exercise 1 – backwards-forwards over a pole/basic (page 178).

IX. Exercises to train body awareness and co-ordination of the hind limbs
Exercise Band 3: Exercise 2 – sit-stand on wobbly back foot target/advanced (page 181).

7. Back relaxation

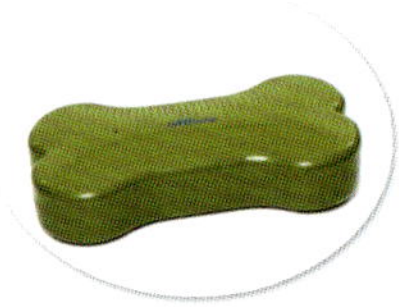

▸▸ **VI. Exercises to relax the back muscles**
Exercise Band 1: Exercise 2 ¬ plank with lowered head/advanced (page 136).

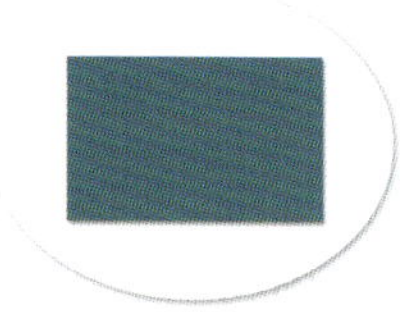

▸▸ **VI. Exercises to relax the back muscles**
Exercise Band 5: Exercise 1 – look under the belly/basic (page 144).

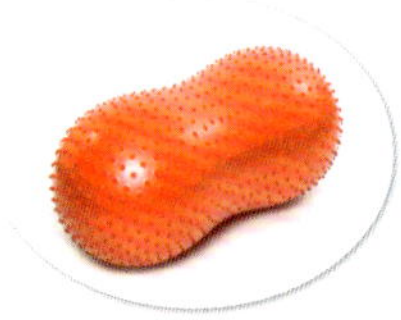

▸▸ **VI. Exercises to relax the back muscles**
Exercise Band 2: Exercise 2 – lying side-long over peanut ball/advanced (page 139).

8. Hindquarters quadriceps group

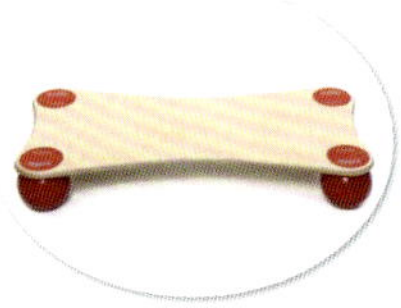

▸▸ **IV. Exercises for the front thigh muscles**
Exercise Band 1: Exercise 1 – sit-stand-sit/basic (page 95), then
Exercise 2 – sit-stand-sit, front elevated/advanced (page 96).

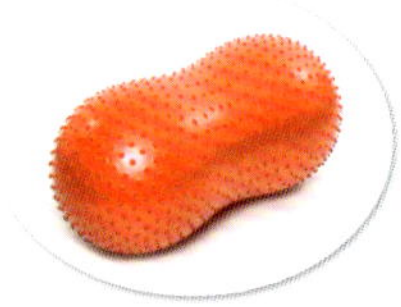

▸▸ **IV. Exercises for the front thigh muscles**
Exercise Band 1: Exercise 4 – knee bend on a peanut ball/professional (page 98)
then Exercise 6 – kick back on a peanut ball/professional (page 101).

▸▸ **IV. Exercises for the front thigh muscles**
Exercise band 6: Exercise 1 – weight shift on a wobble board/basic (page110).
Exercise Band 7: Exercise 2 – bounce to a hand touch /advanced (page 111) x three.

▸▸ **Exercise Band 2:** Exercise 2 – mountain goat/wobbly/advanced (page 114).

9. Fore limbs

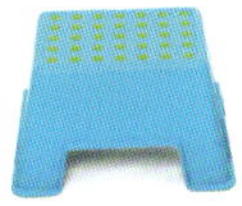

III. Exercises for the fore limbs
Exercise Band 7: Exercise 1 – weight shift to the side/basic (page 83) , then Exercise 2 – weight shift to the side, front elevated /advanced (page 84).

III. Exercises for the fore limbs
Exercise Band 9: Exercise 2 – push-ups/wobbly/advanced (page 89).
Exercise Band 3: Exercise 1– elephant trick with elevated hindquarters/basic (page 175).

10. Sideways track

V. Exercises for the outer thigh muscles
Exercise Band 5: Exercise 1 – side-stepping with front and rear limbs stationed/basic modified (page 130) – change three times laterally, from fitbone to fitbone.

V. Exercises for the outer thigh muscles
Exercise Band 5: Exercise 1 – side-stepping with front and rear limbs stationed/basic (page 130).

V. Exercises for the outer thigh muscles
Exercise Band 1: Exercise 3 – stepping sideways over a series of poles/ professional (page 120).

Cavaletti training

Cavaletti training originally comes from the horse world where it is used for gait training. Cavaletti training is suitable for every dog and can be used in a variety of different ways. You can use a single pole and teach the dog to step sideways, forwards or backwards, or work with a series of poles at different heights. Cavaletti training improves joint mobility, promotes a physiological gait pattern and trains co-ordination.

You can work with the following options:

Spatial training

First determine the optimum distance between the poles. You can do this by measuring from the top of the dog's shoulder blade to the ground, and then setting out the poles at a corresponding distance.

Since you want to achieve a forward orientation, take two floor targets, or similar, and let the dog move from target to target, without a pole. Initially, place both targets very close together, and reward for calm, focused engagement.

If the dog runs at a steady trotting speed from target A to target B, insert a pole in the middle, then two, then four, and finally six poles. The height of the poles should not exceed the height of the dog's hock (lowest rung of the cone). Ideally, the dog should move at a loose trot over the poles, and never jump.

Strength training

In order to build strength, you want to encourage the dog to lift his feet as high as possible. To achieve this, you need to place the poles very close together – about a hand's width between the cones – and at a height that is equivalent to the distance between the dog's ankle and the bend of the knee. This set-up allows the dog to move in a very controlled fashion. However, it is very strenuous and should not be repeated too often.

Different figures (step tempo)

- Mikado – poles lie criss-cross; the dog must concentrate very hard to navigate this.
- Staircase – this incorporates a bend in the spine.
- Star shape – the dog needs to bend to navigate this.
- Lying 8 – requires a bend on both sides.
- Square – allows for different variations.
- Zig zag – the dog needs to calculate different distances.
- Over and under – the dog steps over one pole and ducks under the next.
- Regular and varying distances – promotes concentration.
- Backwards over cavaletti – very strenuous.